Encyclopaedia of
Nuclear Physics

Encyclopaedia of Nuclear Physics

Volume 1

Nuclear Physics

Shahzad Ahmad

ANMOL PUBLICATIONS PVT. LTD.
NEW DELHI - 110 002 (INDIA)

ANMOL PUBLICATIONS PVT. LTD.
H.O.: 4374/4B, Ansari Road, Daryaganj
New Delhi - 110 002
Ph.: 23261597, 23278000
B.O.: No. 1015, Ist Main Road, BSK III Stage
III Phase, III Block
Bangalore - 560 085 (India)
Visit us at: www.anmolpublications.com

Encyclopaedia of Nuclear Physics

First Published, 2006

ISBN 81-261-2618-3 (Set)

PRINTED IN INDIA

Published by J.L. Kumar for Anmol Publications Pvt. Ltd., New Delhi - 110 002 and Printed at Mehra Offset Press, Delhi.

Contents

Preface

Science is the sermon of today. In every walk of life, since childhood till old age, science plays an important role in the life of man. Uptill some decades back, studied merely as a subject in schools and colleges, Science has assumed the status of an essential part of life today. With the expansion of literacy among the masses and growing awareness, they have woken up from the slumber of ignorance and have started taking up things more sincerely and thoughtfully. It is one of the reasons that our nation is heading towards new goals day after day.

It is a vast subject to study and talk about. For the purpose of ease, Science has been divided into more than two hundred and fifty branches. Specialisation in any branch can be obtained at different institutions and universities, scattered all over the country and abroad.

Apart from other super-specific branches, Physics, Chemistry, Zoology, Botany, Mathematics and Computronics, etc. are more general and closer to daily human life. From the point of view of the cut-throat competition in educational world, a thorough study of the above-mentioned branches, proves very handy, for the new aspirants, especially and for teachers and tutors, generally.

An urgent need was being felt for such a series of books, encompassing the commonly-studied and discussed branches of Science of equal significance and presented in an easy language, understood by all. With God's grace, our dream came true and we are happy, while introducing the dream-series for the lovers of knowledge.

The series consists of many a title, covering various branches of Science, such as Physics, Chemistry, Zoology, Botany, Geology, Agriculture Science, Computer Science, Mathematics and so on. The series is important and a must not only for students but also for teachers. Explained in simple language and aided with graphics and diagrams, everything, between the two ends—title and back cover—of the series, has been presented in a perfect way.

This set of books is devoted to **Nuclear Physics**, which comprises three volumes: (1) Nuclear Physics, (2) Atomic Physics and (3) Quantum Physics.

It is hoped that the series would serve both, the students community and the academics, equally and will prove to be a boon for those, aspiring for newer heights. Nevertheless, any suggestions for further betterment of the endeavour will be highly appreciated and solicited.

Editor

1

Radioactive Split

Nuclear Fission

In 1939, two German scientists Hahn and Strassman observed that when uranium nucleus (Z=92) was bombarded with neutrons, it splitted up into two radioactive nuclei which were identified as isotopes of barium (Z=56) and krypton (Z=36). Frisch and Meitner called this phenomenon as 'nuclear fission'. Thus, *nuclear fission is a process in which a heavy nucleus, after capturing a neutron, splits up into two lighter nuclei of comparable masses.* The product nuclei are called 'fission fragments'. The process is accompanied by the release of a few fast neutrons and a huge amount of energy in the form of the kinetic energy of the fission fragments, and also as γ-rays (Fig.).

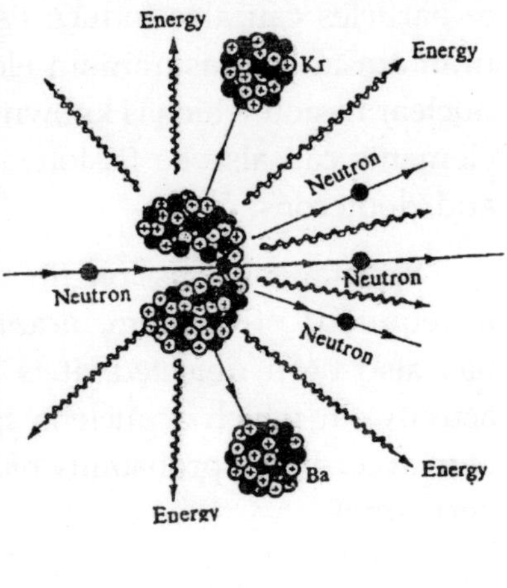

The natural uranium is a mixture of two isotopes, $_{92}U^{238}$ and $_{92}U^{235}$ in the ratio 145:1. The isotope $_{92}U^{238}$ can be fissioned only by the 'fast' neutrons having energy above 1 MeV, while $_{92}U^{235}$ can be fissioned by quite slow neutrons also, say by thermal (0.03 eV-energy) neutrons. Thus when a slow neutron strikes $_{92}U^{235}$ nucleus, it is captured and a highly unstable nucleus $_{92}U^{236}$ is formed which at once breaks up into two fragments with the emission of two or three fast neutrons. A wide range of fission fragments is possible. One of the typical fission reactions is

$$_{92}U^{235} + {}_0n^1 \rightarrow ({}_{92}U^{236}) \rightarrow {}_{56}Ba^{144} + {}_{36}Kr^{89} + 3\ {}_0n^1.$$

In general, the fission fragments are found to be the radioactive isotopes of elements lying in the mass number range roughly from 70 to 160. All fragments undergo several β–decays (electron emissions) until they reach some stable end-product.

Later researches showed that besides uranium, other nuclei are also assignable. Thorium ($_{90}Th^{232}$) and protactinium ($_{91}Pa^{231}$) can be fissioned by fast neutrons, whereas the transuranium element plutonium ($_{94}Pu^{239}$) and an artificial isotope of uranium ($_{92}U^{238}$) can be fissioned by thermal neutrons.

Besides neutrons, accelerated protons, deuterons and α–particles can also induce fission in the nuclei of thorium, uranium and transuranium elements. Even γ-rays can cause nuclear fission which is known as 'photo-fission . Some lighter elements can also be fissioned by very high energy photons and deuterons.

'Spontaneous' fission (in which no bombarding particle is required) of thorium, uranium and transuranic elements has also been detected. It is a process (like natural radio-activity) in which a nucleus splits into two fragments of its own accord. The probability of this type of fission is, however, very small.

Bohr-Wheeler Theory of Nuclear Fission : We know that U^{238} can be fissioned by fast neutrons only (having energy 1.0 MeV or more), while U^{235} is fissionable by fast as well as by slow neutrons. An explanation of this observation was given by Bohr and Wheeler in terms of the liquid-drop model of the nucleus.

A nucleus is like a spherical drop of an incompressible liquid which is held in equilibrium by a balance between the short-range, attractive forces between the nucleons and the repulsive electrostatic forces between the protons. The internucleon forces also give rise to surface tension forces which maintain the "spherical" shape of the drop (Fig. a).

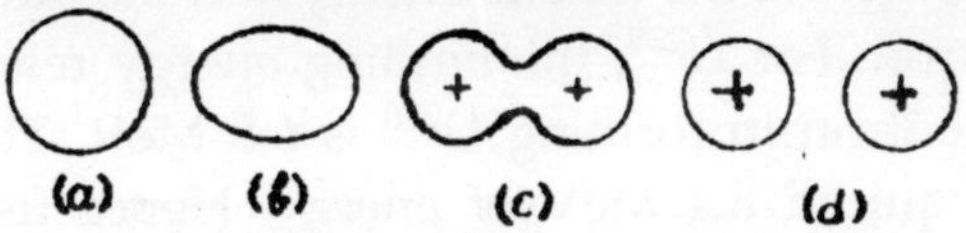

When the nucleus-drop captures a neutron, it becomes a compound nucleus of very high energy. The energy added to the nucleus is partly the kinetic energy of the incident neutron and partly the binding energy of the same neutron which it releases on being captured. This excitation energy initiate rapid oscillations within the drop which at times become ellipsoidal in shape (Fig. b). The surface tension forces tend to make the drop return to its spherical shape, while the excitation energy tends to distort the shape still further. If the excitation energy is small, the nucleus oscillates until it eventually loses its excitation energy by γ-emission (radiative capture) and returns to the spherical shape. If the excitation energy is sufficiently large, however, the nucleus may attain the shape of a dumbbell (Fig. c). When this happens, the repulsive electrostatic forces between the two parts of the dumbbell overcome the attractive forces between nucleons and the splitting takes place. Each splitted part becomes spherical in shape (Fig. d). Thus there is a threshold *activation* energy required to produce stage (c) after which the compound nucleus is bound to split.

The activation energy required to induce fission in various heavy nuclei can be calculated from the liquid-drop model. It gives following values for the Uranium nuclei :

Target Nucleus	Compound Nucleus	Activation Energy
$_{92}U^{235}$	$[_{92}U^{236}]$	6.4 MeV
$_{92}U^{238}$	$[_{92}U^{239}]$	6.6 MeV

When a neutron is captured by one of these nuclei, an energy equal to the corresponding activation energy must be supplied if fission is to occur. The actual energy supplied is equal to the sum of the kinetic energy and the binding energy of the neutron. For U^{235}, the binding energy released by the captured neutron in forming U^{236} is 6.8 MeV. This is *greater* than the required 6.4 MeV of energy. Hence fission would occur even if the kinetic energy of the incident neutron is near zero. This is why the thermal neutrons (energy 0.03 eV) cause fission in U^{235}.

For U^{238}, however, the binding energy released by the captured neutron in U^{239} is only 5.5 MeV which is *less* than the required 6.6 MeV by 1.1 MeV. The incident neutron must therefore have a kinetic energy of atleast 1.1 MeV to induce fission. This is the reason that U^{238} can be fissioned by fast neutrons only, with energy above 1 MeV.

For the same reason the nuclei $_{92}U^{233}$ and $_{94}Pu^{239}$ are fissionable by slow neutrons, while $_{90}Th^{232}$ and $_{91}Pa^{231}$ can be fissioned by fast neutrons only.

Fission Fragments : In a given nucleus the fission may occur in a number of different ways. In general, the fissionable nucleus gives *two* fission fragments which thereafter decay by β–emission into stable end-products. What particular fragments are produced by the given nucleus is a matter of chance. For example, in case of the fission of U^{235} by thermal neutrons, a wide-range of primary fission-fragments having mass numbers roughly between 70 (Fig.) and 160 is possible. The mass distribution of the fission fragments may be shown by a 'fission yield curve', in which the percentage yields of the different fragments are plotted against mass number (Fig.). The curve shows that the splitting of U^{235} nucleus into two fragments of equal mass (A ≈118) has only a 0.01% chance of occurring; whereas the formation of fragments with mass numbers 95 and 140 is most likely (7%). About 97% of the U^{235} nuclei undergoing fission give fragments which fall into two groups, a lighter group with mass numbers from 85 to 104, and a heavier group with mass numbers from 130 to 149.

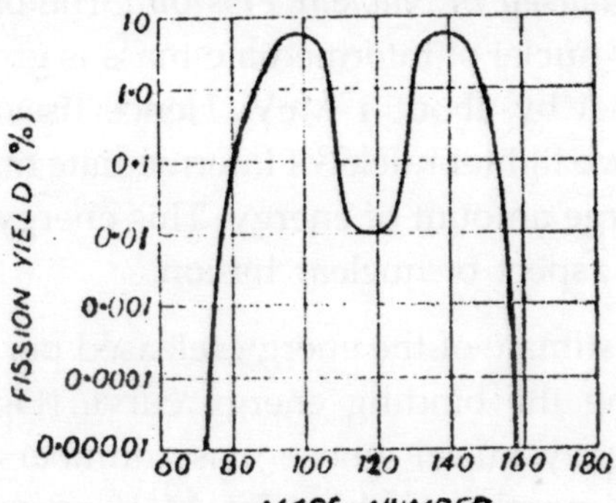

Neutron Emission in Nuclear Fission : An important feature of nuclear fission is the emission of fast neutrons. Most of them (more than 99%) are emitted almost instantaneously (≈ within 10^{-14} sec) with the fission process. These are called 'prompt' neutrons. They have an energy distribution, with an average energy of 2 MeV. In addition, a few neutrons (less than 1% of the total) are emitted a short time later (ranging from 0.05 sec to 1 min) after the fission has occurred. These are called 'delayed' neutrons. On the average, 2.5 neutrons are emitted per fission.

The reason of the emission of the prompt neutrons is as follows. The heavy nuclei have a greater neutron/photon ratio than the medium-mass nuclei. Therefore when a heavy nucleus splits into lighter nuclei, the primary fission fragments have a n/p ratio which is too large for their stability. In other words, the fragments are overloaded with neutrons. Hence they liberate few neutrons as soon as they are formed. These are the prompt neutrons.

The liberation of prompt neutrons does not completely eliminate the neutron-overloading of fission fragments. Therefore as a further means of decreasing n/p ratio, the fragments undergo a chain of β–decays attended by the emission of γ-rays until stable end-products are reached. In a few cases neutrons are emitted

Energy Released in Nuclear Fission : The binding energy *per nucleon* for nuclei of intermediate mass is greater than that of heavy nuclei by about 1 MeV. Hence fission of a heavy nucleus into two lighter nuclei of intermediate masses is bound to release a large amount of energy. This energy release is the most striking aspect of nuclear fission.

A rough estimate of the energy released per fission can be made by using the binding energy curve (Fig.). The curve shows that heavy nuclei whose mass numbers are near 240 have binding energies of about 7.6 MeV per nucleon; while fission fragments whose mass numbers lie in the range roughly from 70 to 160 have binding energies of about 8.5 MeV per nucleon. Thus the fission fragments have, on the average, binding energy of about 0.9 MeV per nucleon greater than the (heavy) nucleus which has been fissioned. Hence 0.9 MeV/ nucleon is released during the fission. The total amount of energy released per fission of a compound nucleus U^{236} (which has 236 nucleons) is thus

$$(0.9 \text{ MeV}) \times 236 = 212 \text{ MeV},$$

or roughly 200 MeV. This is much larger than the energy liberated in nuclear reactions (other than fission) which is of the order of 10 MeV. (The energy liberated in an ordinary chemical reaction is only a few electron volts. The energy released per fission is distributed roughly as follows :

Fission fragments (kinetic energy)	168 MeV
Prompt neutrons (kinetic energy)	5 MeV
Prompt γ-rays	5 MeV
β–decay energy from fission fragments	5 MeV
γ-rays from fission fragments	7 MeV
Neutrino energy from β–decay	10 MeV
	200 MeV

Thus we see that most of the (84%) energy released during fission goes into the kinetic energy of the fission fragments, 5% is given off as kinetic energy of neutrons and γ-rays which are emitted at the time of fission, and the remainder (11%) appears as radioactivity of the fission, fragments which decay to form stable end-products.

Chain-reaction : When a neutron fissions a uranium nucleus then, besides the fission fragments, a few fast neutrons are also emitted. If one or more of the emitted neutrons are used to fission other nuclei, further neutrons are produced and the process is repeated. The reaction thus becomes self-sustained and is known as a chain reaction (Fig.).

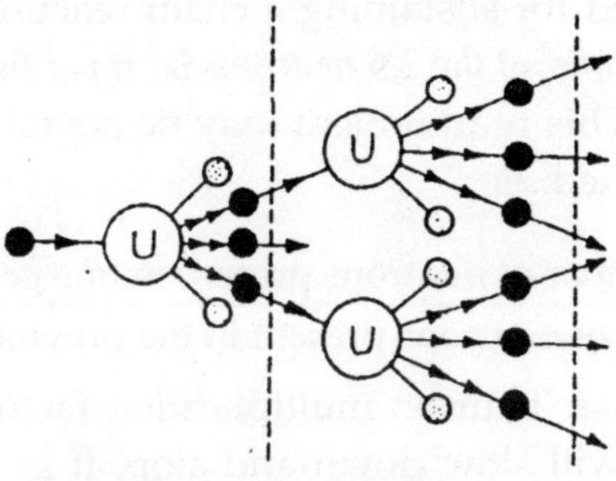

The chain reactions may be of two types :

(i) *Uncontrolled Chain Reaction :* If more than one of the neutrons emitted in a particular fission cause further fissions, then the number of fissions increases rapidly with time. Since a large amount of energy is liberated in each fission, within a very short time the energy takes a tremendous magnitude and is released as a violent explosion. This is what happens in a nuclear bomb. Such a chain reaction is called an "uncontrolled" reaction.

(ii) *Controlled Chain Reaction :* If by some means, the reaction is controlled in such a way that only one of the neutrons emitted in a fission causes another fission, then the fission rate remains constant and the energy is released steadily. Such a chain reaction is called a "controlled" reaction. It is used in a nuclear reactor.

The two types of chain reactions take place in different situations.

General Condition for a Self-Sustained Chain Reaction—Critical Size :In the fission of uranium nuclei, 2.5 neutron are, on the average, emitted per fission. Not all of these neutrons are available for further fissions. Some of them escape through the surface of the uranium, while many are lost in non-fission processes such as radiative capture by the nuclei of uranium and other materials present. Hence the basic condition that must be satisfied for sustaining a chain reaction is that *on the average, at least 'one' of the 2.5 neutrons born per fission must cause another fission.* This requirement may be stated by means of a constant k defined as :

$$k = \frac{\text{number of neutrons present in one generations}}{\text{number of neutrons present in the previous generation}}$$

k is known as 'neutron multiplication factor'. If $k < 1$, the chain reaction will slow down and stop. If $k-1$, the reaction will proceed at a steady rate. If $k > 1$, however, the reaction will grow to an explosive rate. These situations are known as 'sub-critical' 'critical' and 'super-critical' respectively.

In a nuclear bomb, values of k considerably greater than 1 are needed ; while in a nuclear reactor, k is required to be 1 or only slightly greater than 1.

In order to have the neutron multiplication factor greater than 1, the rate of production of neutrons must be larger than the rate of their loss. If the uranium is in the form of a solid sample, the rate of production of neutrons will be proportional to its volume while the rate of their escape will be proportional to the surface area. Since for a given volume, the sphere has the smallest surface area, the uranium should be taken in the form of a sphere.

Now, let us consider a uranium sphere of radius R. Let N_1 be the number of neutrons produced in a given time-interval, N_2 the number of neutrons lost in non-fission processes and N_3 the number escaped through the surface in the same time-interval. N_1 and N_2 will be proportional to the volume, while N_3 will be proportional to the surface area of the sphere. Thus

$$N_1 \propto \frac{4}{3}\pi R^3 = C_1 R^3,$$

$$N_2 \propto \frac{4}{3}\pi R^3 = C_2 R^3,$$

and $$N_3 \propto 4\pi R^2 = C_3 R^2,$$

where C_1, C_2 and C_3 are proportionality constants. For $K > 1$ we must have

$$N_1 > N_2 + N_3$$

or $$C_1 R^3 > C_2 R^3 + C_3 R^2$$

or $$C_1 R > C_2 R + C_3$$

or $$(C_1 - C_2) R > C_3$$

or $$R > \frac{C_3}{C_1 - C_2} = C \text{ (say)}.$$

C is known as the critical size of the sample. Thus, *in order to achieve a self-sustained chain reaction, the size (or mass) of the sample must be greater than a critical value* C. Below the critical size (or critical mass) the chain reaction would not occur.

Further, even if we take a piece of natural uranium in a size greater than the critical value, the chain reaction would not develop in it. The reason is that the neutrons emitted in a fission carry an energy of about 2 MeV. Such 'fast neutrons' have a much larger probability of being scattered by U^{238} than fissioning it. In this process their energy is reduced to below 1 MeV and then they are captured by U^{238} without fission. U^{235} nuclei, however, can be fissioned by fast neutrons but they are insignificantly small in number (only 0.7%) in natural uranium.

The isotope U^{235} can, however, be separated from natural uranium and then used as fissionable material. It can be fissioned by neutrons of *all* energies and has a much larger probability of fission than radiative capture. Hence, once a fission is initiated, the chain reaction is built up at an explosive rate.

Achievement of Controlled Chain Reaction : A different way of carrying out a chain reaction in 'natural' uranium is to rapidly slow down the emitted neutrons to thermal energies ($\approx$ 0.03 eV) by means of moderators. The fission probability of U^{235} by thermal neutrons is very large. Therefore, inspite of the very small concentration of U^{235} in natural uranium, fissions of U^{235} occur slightly more frequently than the radiative (non-fission) capture of these neutrons both by U^{235} and U^{238}. Hence the neutron reproduction factor becomes very slightly greater than 1 and the chain reaction continues at almost a steady rate. This method is employed in a nuclear reactor.

Function of Moderator

A controlled chain-reaction is obtained in a nuclear reactor by slowing down the neutrons emitted in nuclear fissions. This is achieved by colliding the (fast) neutrons with such nuclei which have a small probability for neutron capture but a large probability for neutron scattering. A substance whose nuclei have these properties is called a 'moderator'. The nuclei of the moderator absorb energy from the colliding neutrons which are then scattered with reduced energy. The energy-loss is maximum if the nuclei are of the same mass as neutrons. Heavy water (D_2O) which has heavy-hydrogen nuclei, and pure graphite which has carbon nuclei are the best moderators. About 250 collisions are required to reduce the energy of a neutron from 2 MeV to thermal energy in heavy water and about 100 collisions in graphite. When heavy water is the moderator, the uranium is taken in the form of very small particles of uranium oxide suspended in the (heavy) water. The reactor using it is said to be 'homogeneous'. When graphite is the moderator, then uranium is taken in the form of rods distributed through the graphite in the form of a lattice. The reactor using it is said to be 'heterogeneous'.

Pu^{239} versus U^{235} as Fissionable Material : Pu^{239} has almost same fission properties as U^{235}. Hence both can be used for fissioning. But, however, the separation of U^{235} from natural uranium (99.3% U^{238} + 0.7% U^{235}) is not easy. The reason is that U^{235} and U^{238} have same chemical properties and so they cannot be separated from each other by any chemical method. They can be separated either by thermal diffusion or in a mass spectrograph. These processes are, however, very slow and expensive because of a very small fractional difference in the masses of U^{235} and U^{238} and the very small relative concentration of U^{235}.

Pu^{239} is produced by neutron bombardment of U^{238} and is a chemically different element. Therefore it can be easily separated from uranium by chemical methods. Hence Pu^{239} is now being used all over in nuclear bombs and nuclear reactors.

Atom Bomb

An atom bomb, infact a nuclear fission bomb, is a device in which an *uncontrolled* chain reaction is built up in a fissionable material by means of fast neutrons. It then releases tremendous amount of energy in a very short time.

The bomb consists essentially of two pieces of U^{235} (or Pu^{239}), which are kept separated in a massive cover of a high-density material (Fig.). The mass of each piece is below the critical value so that any stay neutron, either from cosmic rays or produced by spontaneous fission, is unable to start a chain reaction. Thus, so long the two pieces are kept separated, they are perfectly stable and safe.

When an explosion is required, the two pieces are brought rapidly together (in a time less than a microsecond) by means of some device so that the total mass becomes supercritical *(greater than the critical value)*. As soon as this happens, the stray neutrons initiate chain reaction in which fast neutrons rapidly multiply and within a few millionths of a second the reaction acquires an explosive nature.

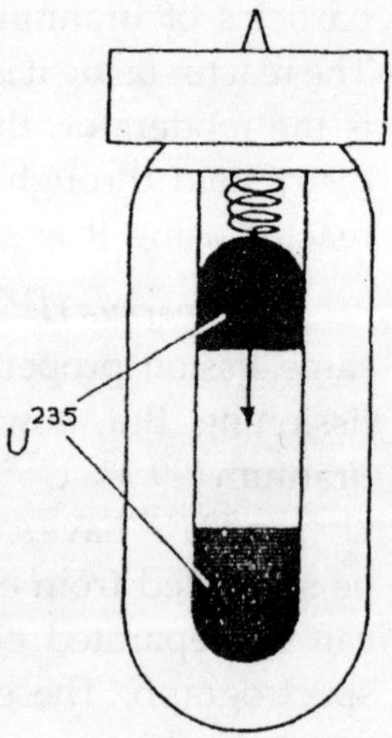

An idea of the rate of energy released in a nuclear bomb having 1 kg of uranium can be made in the following way :

In a nuclear chain reaction the number of neutron N, available to produce fission at any time t is given by

$$N = N_0 e(k - 1)\ t/\tau$$

where k is the neutron multiplication factor and τ is the mean life-time of fission neutrons, τ is the time which elapses between the emission of a fission neutron and the production by this neutron of a further fission neutron. The average value of τ in a nuclear bomb is one nano-second (10^{-9} sec). Therefore the increase in the neutron population after 1 microsecond ($t = 10^{-6}$ sec) is given by (taking k =1.1).

$$\frac{N}{N_0} = e^{0.1} \times 10^{-6}/10^{-9} = e^{100} \simeq 10^{43}.$$

Thus within a micro-second the neutrons multiply many millions of times which are sufficient *to* fission all the 10^{24} uranium nuclei present in 1 kg. Since each fission produces about 200 MeV of energy, an amount of 2×10^{26} MeV of energy could be released within 1 micro-second. This will constitute a large explosion.

In the explosion, a temperature of the order of 10^7 °C and a pressure of several million atmospheres is developed. Large quantities of radioactive material and blinding flashes of light are also produced. All objects and living creatures within a range of hundreds of kilometers are completely destroyed. The radioactive dust formed is carried away by air-currents to distant places causing loss beyond description. The bomb dropped on Hiroshima in 1945 had released an amount of energy equivalent to that from 20,000 tons of TNT. It was called a 20-kiloton bomb.

An upper limit is set to the size of a fission bomb by the practical difficulty of combining more than two pieces at exactly the same time, and because each of the pieces must be below the critical size so that it can be transported safely.

Nuclear Reactor

A nuclear reactor, or a nuclear pile, is a device in which a self-sustaining, *controlled* chain reaction is produced in a fissionable material. It is thus a source of controlled energy. The main parts of a modern reactor are as follows :

(i) *Fuel :* The fissionable material, known as fuel, plays the key role in the operation of a reactor. Uranium enriched with the isotope U^{235}, or Pu^{239}, is used as fuel.

(ii) *Moderator :* The moderator slows down the neutrons to thermal energies by elastic collisions between its nuclei and the fission neutrons. The thermal neutrons have a very high probability of producing fission in U^{335} nuclei. Heavy water, graphite and beryllium oxide are most suitable moderators.

(iii) *Control Rods:*These are the rods of cadmium (or boron) which are used to control the fission rate in the reactor. Cadmium and boron are good absorbers of slow neutrons. Therefore, when the rods are pushed into the reactor, the fission rate decreases, and when they are pulled out, the fission grows.

(iv) *Shield :* Since nuclear fissions produce various types of radiations which are dangerous to human life, the reactor is surrounded by a concrete wall about 2 meter thick and containing a high proportion of elements like iron.

(v) *Coolant :* The energy is released inside the reactor in the form of heat which is removed by means of a cooling agent, known as coolant. The carbon dioxide gas is the main coolant in a reactor. It is circulated through the interior of the reactor by a pumping system.

(vi) *Safety Device :* In an emergency, if the reactor begins to go too fast, a special set of control rods, known as shut-off rods, drop in automatically. They immediately absorb the neutrons so that the chain reaction stops entirely.

Construction : One type of nuclear reactor is shown in Fig. It consists of a large number of uranium rods placed in a calculated geometrical lattice between layers of pure graphite (moderator) blocks. To prevent oxidation of uranium and also to preserve the gaseous fission products, the rods are covered by close-fitting aluminium cylinders. The control rods are so inserted in the lattice that they can be raised or lowered between the uranium rods whenever necessary. The whole reactor is surrounded by a concrete shield.

Working : The actual operation of the reactor is started by raising the control rods so that they do not absorb many neutrons. Even a single neutron is capable of starting fission and there are always a few stray neutrons present either from the cosmic radiation or from a spontaneous fission. As soon as a neutron strikes a U^{235} nucleus and fissions it, two or three fast neutrons are emitted. These neutrons are slowed down from energies of several MeV to energies of less than 1 eV by collision with moderator nuclei, after which they induce further fission in U^{235}. The reaction, once started, is controlled by moving the control rods in and out.

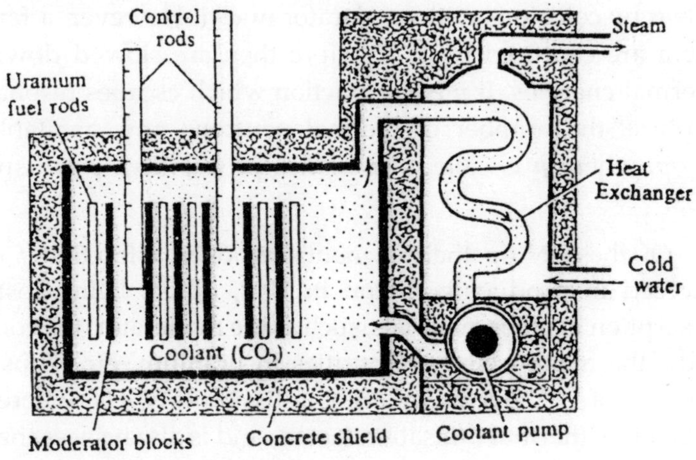

Carbon dioxide is pumped rapidly through the reactor to carry away the heat generated by the fission of the uranium nuclei. The hot carbon dioxide gas passes through a heat-exchanger where it gives up its heat to water and converts it into steam. This steam drives the turbines and generates electric power.

The size of the reactor may be reduced by using heavy water as a moderator (in place of graphite). It, being lighter, slows the neutrons more effectively.

Neutron Cycle : The neutron balance in a thermal reactor can be described in terms of a cycle. For simplicity we assume that the reactor is infinitely large so that there is no leakage of neutrons through its surface. Let us start with the fission of U^{235} nucleus by a thermal neutron. In this process, suppose N_0 fast neutrons are emitted. These neutrons have an energy above the fission threshold for U^{238} and so some of these may produce fission in U^{238} before colliding with moderator nuclei, causing a fractional increase $\in$. The number of fast neutrons available is now $N_0\in$ where $\in$ is called the fast fission factor', $\in$ is usually about 1.03.

The $N_0\in$ neutrons diffuse through the pile and are slowed down by collisions with moderator nuclei. However, a few of them are captured by U^{288} before they are slowed down to thermal energies. If p is the fraction which escapes resonance capture, the number of thermal neutrons now available is $N_0\in p$, where p is called 'resonance probability'. p is usually 0.95.

Of these $N_0\in p$ thermal neutrons, only a fraction f may succeed in producing fission in U^{235}, others being lost by absorption in other materials, such as the moderator, the control rods, the metal casing, impurities in uranium, etc. Thus the number of U^{235} nuclei undergoing fission is $N_0\in pf$, where f is called the 'thermal utilisation factor', and is always less than 1.

If each thermal fission in U^{235} produces η 'fast' neutrons to start the cycle over again, then the total number of neutrons after one cycle, or one generation, is

$$N = N_0 \epsilon p f \eta.$$

The multiplication factor for a reactor (of infinite dimensions) is therefore

$$k = \frac{N}{N_0} = \epsilon p f \eta.$$

This formula is called the 'four-factor formula.'

For the steady operation of the reactor at a particular power level, k must be equal to 1. The reactor is then said to be 'critical. The condition is achieved by varying the factor f by adjustment of control rods. If $k < 1$ then the chain reaction stops and the reactor becomes sub-critical. If $k > 1$, the fissions increase from cycle to cycle and the reactor becomes supercritical.

Role of Delayed Neutrons

While most of the fission neutrons are emitted promptly, about 0.7% are emitted a short time later after the fission has occurred. These delayed neutrons help in increasing the over-all average life-time of fission neutrons. The average life-time of the prompt neutrons in a graphite reactor is about 1 milli-second (10^{-3} sec) but with the inclusion of the delayed neutrons the over-all average increases to 0.1 second. Thus sufficient time is available for adjusting the control rods. If *all* the neutrons were prompt, the reactor would become supercritical and would be unmanageable.

Applications **:** The nuclear reactors are used mainly for the following purposes :

(i) *Production of Pu^{239}:* The ordinary uranium reactor is used to produce plutonium (Pu^{239}) which is a better fissionable material than U^{235}. The neutrons in the reactor which do not participate in the fission chain-reaction of U^{235}, are absorbed by U^{238} and convert it into a heavier isotope U^{239} :

$$_{92}U^{238} + {_0}n^1 \rightarrow {_{92}}U^{239} + \gamma \text{ (energy)}$$

$_{92}U^{239}$ is an unstable nucleus. It emits a β–particle and is converted into a new heavier element, neptunium ($_{93}Na^{239}$) :

$$_{92}U^{239} \rightarrow {_{93}}Np^{239} + {_{-1}}\beta^0 + \bar{v}$$

(anti-neutrino)

Neptunium also emits a β–particle and is converted into plutonium ($_{94}Pu^{239}$) :

$$_{93}NP^{239} \rightarrow {_{94}}Pu^{239} + {_{-1}}\beta^0 + \bar{v}.$$

Since plutonium is a new element and is different in chemical properties from uranium, so it can easily be separated from uranium. It is used in nuclear reactors and in nuclear bombs in place of U^{235}. The critical size of Pu^{239} is smaller compared to that of U^{235} for the fission chain-reaction. Therefore, the use of Pu^{239} is economical compared to U^{235}.

(ii) *Production of Neutron Beam :* Fast-moving neutrons are emitted by the fission of U^{235} in the reactor. By converging these neutrons into a fine beam, artificial disintegration of other elements are studied.

(iii) *Production of Radio-Isotopes :* Artificial radioactive isotopes of many elements are produced in the reactor. For this, the element is placed in the reactor and bombarded with fast-moving neutrons. These radio-isotopes are utilised in medicines, biology, agriculture, industries and scientific discoveries.

(iv) *Generation of Energy :* The energy produced in reactors is used to run electric-generators. Thus nuclear energy is converted into electrical energy. It is used at power stations to generate electricity on a large scale which runs industries. Nuclear energy can be used in place of coal and patrol for driving the engines and for the propulsion of ships, submarines and aircrafts.

Thermal Reactor and Breeder Reactor : The reactors in which energy is produced by the fission of U^{235} by slow neutrons, are called 'thermal reactors'. Since the major part in ordinary uranium is U^{238} (U^{235} is only 0.7%), therefore the fission of U^{235} is very costly, and this would lead to an early depletion of uranium reserves. We know that besides U^{235}, Pu^{239} and U^{238} are also fissionable substances. So Pu^{239} is produced from U^{238} in many nuclear reactors. In addition to it, U^{233} is also produced from Th^{232}. The quantity of the substance produced (Pu^{239} and U^{233}) in these reactors is more than the quantity of substances (U^{238} and Th^{232}) consumed. Such reactors are called breeder reactors'.

Nuclear Fusion

When two or more very light nuclei moving at very high speeds are fused together to form a single nucleus, then this process is known as 'nuclear fusion'. The mass of the product nucleus is less than the sum of the masses of the nuclei which are fused. The lost mass is converted into energy which is released in the process. This property of the light nuclei is shown by the binding energy curve (Fig.), in which the average binding energy per nucleon rises rapidly with increase in mass number in the range of low mass number nuclei.

For example, two deuterons (heavy-hydrogen nuclei) can be fused to form a tritron (tritium nucleus) according to the following reaction :

$$_1H^2 + {_1H^2} \rightarrow {_1H^3} + {_1H^1} + 4.0 \text{ MeV energy.}$$

The tritron so formed can further fuse with a third deuteron to form an α–particle (helium nucleus):

$$_1H^3 + {}_1H^2 \rightarrow {}_2He^4 + {}_0n^1 + 17.6 \text{ MeV energy.}$$

The net result of the two reactions is the burning of three deuterons and the formation of an α–particle ($_2He^4$), a neutron ($_0n^1$) and a proton ($_1H^1$). The total energy released is 21.6 MeV, so that the energy released per deuteron burnt is 7.2 MeV. Most of the liberated energy appears as kinetic energies of neutron and proton.

Alternatively, the fusion of three deuterons into an α–particle can take place as follows :

$$_1H^2 + {}_1H^2 \rightarrow {}_2He^3 + {}_0n^1 + 3.3 \text{ MeV energy,}$$

$$_2He^3 + {}_1H^2 \rightarrow {}_2He^4 + {}_1H^1 + 18.3 \text{ MeV energy.}$$

The net result is same as before ; the energy released per deuteron burnt being again 7.2 MeV.

The energy output in a fusion reaction (21.6 MeV) is much less than the energy released in the fission of a U^{235} nucleus which is about 200 MeV, But this does not mean that fusion is a weaker energy-source than fission. The number of deuterons in 1 gram of heavy hydrogen is much larger than the number of U^{235} nuclei in 1 gm of uranium. Therefore, *the energy output per unit mass of the material consumed is much greater in case of the fusion of the light nuclei than in case of the fission of heavy nuclei.*

The fusion process is, however, not easy to carry out. Since the nuclei *to* be fused are positively charged, they would repel one another strongly. Hence they must be brought very close together not only by high pressure but also with high kinetic energies (≈ 0.1 MeV). For this, a temperature of the order of 10^8 Kelvin is required. Such high temperatures are available in the sun and stars. On earth they may be produced by exploding a nuclear fission bomb. Since very high temperatures are needed for the fusion of nuclei, the process is called a 'thermonuclear reaction', and the energy released is called as 'thermonuclear energy'.

Difference between Nuclear Fission and Nuclear Fusion : These two nuclear processes are opposite in nature but give a common result namely, the liberation of large amount of energy. The fission is the splitting of a 'heavy' nucleus under neutron bombardment into lighter radioactive nuclei whose combined mass is less than the mass of the original nucleus. The lost mass is converted into energy. The energy when released in an uncontrolled manner produces destruction (as in a bomb), but if released steadily it can be used for peaceful purposes (as in a reactor).

The fusion, on the other hand, is the joining of two or more 'light' nuclei into a single nucleus whose mass is less than the sum of the masses of the joining nuclei. Again, the lost mass is converted into energy. This process happens at very high temperature and under very high pressure. The energy released in fusion is uncontrolled, though efforts are being made to control the fusion energy.

Hydrogen Bomb

It is a nuclear fusion bomb (also called as thermonuclear bomb), based upon the fusion of heavy-hydrogen nuclei. Since fusion takes place under the extreme conditions of high pressure and high temperature, a fission bomb must be used as igniter of a fusion bomb.

The central core of a hydrogen bomb is a uranium (or plutonium) fission bomb which is surrounded by the fusable material such as lithium hydride (LiH^2), a compound of heavy hydrogen. Thus, in turn, is surrounded by uranium. When the central fission bomb explodes, enormous temperature and pressure are produced and the fusion of surrounding heavy-hydrogen nuclei starts. The fast neutrons produced in fusion produce further fissions in the outer layer of uranium. Thus a fusion chain reaction is developed and a tremendous amount of energy is released.

The size of a fission bomb cannot be increased beyond a limit because in it the fissionable material is kept in two pieces, and the size of each piece should be less than the critical size. There is no such restriction on the size of hydrogen bomb. The material to be fused may be taken in it in any quantity. Once the fusion is initiated, it can spread throughout any mass of the material. Therefore, a fusion bomb is much more destructive than the fission bomb.

The fusion of nuclei is an uncontrolled process. Till now, there is no available method of controlling the release of fusion energy. Therefore, its only use so far is destructive.

Practical Difficulties in Controlling Thermonuclear (Fusion) Energy : A controlled release of fission energy is possible in a nuclear fission reactor. But there is not yet available any method of controlling the release of fusion energy. Hence a nuclear fusion reactor is so far only a dream.

To carry out controlled nuclear fusion, it is necessary to set up and maintain a temperature of the order of 10^8 K in a limited volume. At such a high temperature the atoms are entirely stripped of their electrons so that the fusable material is a completely ionised gas called 'plasma. The plasma thus consists of atomic nuclei and electrons in rapid random motion. The main problem is the design of a "container" in which the hot plasma can be contained under the required high pressure. No vessel of any substance can be a container because it would immediately evaporate.

Moreover, the contact of plasma with the walls of the vessel will result in its cooling. Attempts are being made to retain the plasma is a circular magnetic field created by means of a current pulse of the order of a million amperes (Fig.). The magnetic forces on the moving charged plasma particles make them travel along paths in a limited part of space. This is known as "pinch effect."

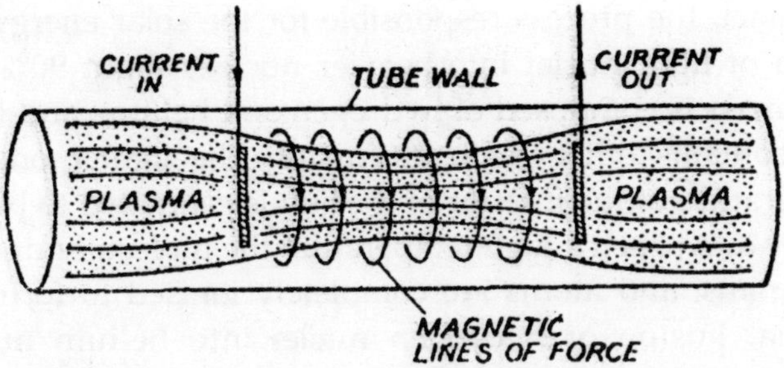

The achievement of controlled fusion will provide mankind with almost limitless source of energy. This is because the main fusion fuel is heavy hydrogen which is found in abundance in the sea water (whereas the fission fuel uranium is limited). Further, the products of fusion are, in general, not radioactive. Therefore, the problem of the disposal of the radioactive waste will not arise.

Solar Energy

The sun has been radiating energy at an enormous rate for billions of years. The source of the continuous supply of energy remained a mystery for a long time. Chemical reactions in the sun cannot supply this energy. The reason is that even if the sun were composed entirely of carbon, then its complete combustion would supply energy at the present rate for only a few thousand years. In that case the sun would have burnt long ago.

Helmholtz suggested that the sun is continuously contracting and so its gravitational energy is being converted into heat energy. But calculations show that contraction could supply not more than one percent of the actual energy output. Hence gravitational contraction can also not be a possible source of solar energy.

Nuclear fission of heavy nuclei must also be ruled out because the abundance of heavy elements in the sun is too small to account for the observed rate of energy emission.

Infact, the process responsible for the solar energy is the fusion of light nuclei into heavier nuclei. About 90% of the solar mass is composed of hydrogen and helium, and the rest 10% contains other elements, mainly the lighter ones. The temperature of the interior of the sun is estimated to be about $2 \times 10^7\ K$. At such a high temperature the molecules dissociate into atoms, and atoms are completely ionised to form a hot plasma. Fusion of hydrogen nuclei into helium nuclei is continuously taking place in this plasma, with the continuous liberation of energy.

It is unlikely in the solar conditions that four hydrogen nuclei would fuse together directly to form a helium nucleus. This may take place through a cycle of processes, however. Two such cycles have been proposed ; the carbon cycle and the proton-proton cycle.

Carbon Cycle : This cycle was proposed by Bethe in 1939 to account for the energy radiated by the sun. In this cycle the fusion of hydrogen nuclei into helium nucleus takes place in the sun through a series of nuclear reactions in which carbon acts as a catalyst. These reactions occur in the following order:

$$_6C^{12} + {_1H^1} \rightarrow {_7N^{13}} + \gamma \text{ (energy)}$$

$$_7N^{13} \rightarrow {_6C^{13}} + {_{+1}\beta^0} + v \text{ (neutrino)}$$

$$_6C^{13} + {_1H^1} \rightarrow {_7N^{14}} + \gamma$$

$$_7N^{14} + {_1H^1} \rightarrow {_8O^{15}} + \gamma$$

$$_8O^{15} \rightarrow {_7N^{15}} + {_{+1}\beta^0} + v$$

$$_7N^{15} + {_1H^1} \rightarrow {_6C^{12}} + {_2He^4}$$

$$\text{on adding : } 4{_1H^1} \rightarrow {_2He^4} + 2\,{_{+1}\beta^0} + 2v + \gamma \text{ (energy)}$$

The net result is the fusion of four hydrogen nuclei into a helium nucleus with the emission of two positrons ($_{+1}\beta^0$) and 24.7 MeV of energy. The nucleus $_6C^{12}$ which starts the cycle reappears in the final reaction. The emitted positrons combine with two electrons and are annihilated, producing about 2

MeV of energy. Thus about 26.7 MeV of energy is released for every helium nucleus formed. Enormous number of such fusions can simultaneously take place in the sun.

Proton-Proton Cycle : Another possible cycle of reactions is the proton-proton cycle which occur in the following sequence:

$$_1H^1 + {_1H^1} \rightarrow {_1H^2} + {_{+1}\beta^0} + v \text{ (neutrino)} \quad \text{... }(i)$$

$$_1H^2 + {_1H^1} \rightarrow {_2He^3} + \gamma \text{ (energy)} \quad \text{... }(ii)$$

$$_2He^2 + {_2He^3} \rightarrow {_2He^4} + {_1H^1} + {_1H^1} \quad \text{... }(iii)$$

Multiplying each of the equations and (ii) by 2, and then adding all the three, we get

$$4\,{_1H^1} \rightarrow {_2He^4} + 2\,{_{+1}\beta^0} + 2v + \gamma \text{ (energy)}$$

In this cycle of reactions, the reactions (*i*) and (*ii*) must occur twice for the reaction (*iii*) to take place. The net result is again the fusion of four hydrogen nuclei into a helium nucleus with the emission of two positrons. Approximately same energy is released per cycle as in the carbon cycle.

In the sun, whose interior temperature is estimated to be 2×10^7 K; the proton-proton cycle has the greater probability for occurrence. In general, the carbon cycle is more efficient at high temperatures, while the proton-proton cycle is more efficient at low temperatures. Hence stars hotter than the sun obtain their energy (known as stellar energy) largely from the carbon cycle, while those cooler than the sun obtain the greater part of their energy from the proton-proton cycle.

The sun is emitting energy at the rate of about 4×10^{26} joule/ second. Because of this, the solar mass is reducing at the rate of about 4×10^9 kg/second. Thus, the sun is annihilating at a very high rate. But the annihilated mass is still very much less than the total mass of the sun which is about 2×10^{30} kg. It is estimated that the sun will be emitting energy at the present rate for the next one thousand crore (10^{11}) years.

PROBLEMS

1. *In a nuclear reactor, fission is produced in 1 gm of U^{235} (235.0439 amu) in 24 hours by a slow neutron (1.0087 amu). Assuming that $_{36}Kr^{92}$ (91.8973 amu) and $_{56}Ba^{141}$ (140.9139 amu) are produced in all reactions and no energy is lost, write the complete reaction and calculate the total energy produced in MeV and in killowatt-hour. Given 1 amu = 931.5 MeV.*

Solution: The nuclear fission reaction is

$$_{92}U^{235} + {}_{0}n^{1} \rightarrow {}_{56}Ba^{141} + {}_{36}Kr^{92} + 3\ {}_{0}n^{1}.$$

The sum of the masses before reaction is

$$235.0439 + 1.0087 = 236.0526 \text{ amu}.$$

The sum of the masses after the reaction is

$$140.9139 + 91.8973 + 3\,(1.0087) = 235.8373 \text{ amu}.$$

The mass loss in the fission is

$$\Delta m = 236.0526 - 235.8373 = 0.2153 \text{ amu}.$$

The energy equivalent of 1 amu is 931.5 MeV. Therefore, the energy released in the fission of a U^{235} nucleus

$$= 0.2153 \times 931.5 = 200 \text{ MeV}.$$

The number of atoms in 235 gm of U^{235} is 6.02×10^{23} (Avogadro number). Therefore, the number of atoms in 1 gm

$$= \frac{6.02 \times 10^{23}}{235} = 2.56 \times 10^{21}.$$

Hence the energy released in the fission of 1 gm of U^{235} *i.e,* in 2.56×10^{21} fissions is

$$E = 200 \times 2.56 \times 10^{21} = 5.12 \times 10^{23} \text{ MeV}.$$

Now, $1 \text{ MeV} = 1.6 \times 10^{-13}$ joule.

$$\therefore \quad E = 5.12 \times 10^{23} \times (1.6 \times 10^{-13})$$

$$= 8.2 \times 10^{10} \text{ joule}.$$

Now, 1 kilowatt-hour= 1000 watt x 3600 sec

$$= 1000 \frac{\text{joule}}{\text{sec}} \times 3600\text{sec}$$

$$= 3.6 \times 10^6 \text{ joules.}$$

$$\therefore \quad E = \frac{8.2 \times 10^{10}}{3.6 \times 10^6} = 2.28 \times 10^4 \text{ kW-H.}$$

2. *Calculate the energy released from the fission of 100 gm of U^{235} if the fission of one U^{235} nucleus gives 200 MeV of energy.*

[Ans: 5.12×10^{25} Mev]

3. *If in a certain fission process, the mass loss is 0.1%, then calculate the energy liberated by the fission of 1 kg of the substance. How much kilowatt-hour electric energy can be generated from it ?*

Solution: The mass loss in 1 kg of the substance is

$$\Delta m = \frac{0.1}{100} \times 1 \text{ kg} = 0.001 \text{ kg.}$$

According to Einstein's mass-energy relation, the energy liberated is

$$\Delta E = (\Delta m)\, c^2$$

$$= 0.001 \text{ kg} \times (3.0 \times 10^8 \text{ m/s})^2$$

$$= 9.0 \times 10^{13} \text{ joule.}$$

Now, 1 kilowatt-hour = 3.6×10^6 joules.

$$\therefore \quad \Delta E = \frac{9.0 \times 10^{13}}{3.6 \times 10^6} = 2.5 \times 10^7 \text{ kilowatt-hour.}$$

4. *Calculate the useful power produced by a reactor of 40% efficiency in which 10^{14} fissions are oçcurring each second and the energy per fission is 250 MeV. Take 1 MeV = 1.6×10^{-13} joule.*

[Ans. 1.6 kilowatts]

5. *Calculate the fission rate of U^{235} required to produce 2 watts and the amount of energy that is released in the complete fission of 0.5 kg of U^{235}. The energy released per fission of U^{235} is 200 MeV.*

Solution: The power to be produced is

$$P = 2 \text{ watts} = 2 \text{ joule/sec}$$

$$= \frac{2}{1.6 \times 10^{-13}} \text{ MeV/sec}$$

$$[\therefore 1 \text{ MeV} = 1.6 \times 10^{-13} \text{ joule}]$$

$$= 1.25 \times 10^{13} \text{ MeV/sec.}$$

The energy released per fission is 200 MeV.

$\therefore$ required no. of fissions per sec

$$= \frac{1.25 \times 10^{13}}{200} = 6.25 \times 10^{10}.$$

The fission rate is 6.25×10^{10} per sec.

The number of atoms in 235 gm of U^{235} is 6.02×10^{23}. Therefore, the number of atoms in 0.5 kg (=500 gm) of U^{235}

$$= \frac{6.02 \times 10^{23}}{235} \times 500 = 1.28 \times 10^{24}.$$

The energy released per fission is 200 MeV. Therefore, the total energy released in the complete fission of 0.5 kg of U^{235}

$$= 200 \text{ MeV} \times (1.28 \times 10^{24})$$

$$= 2.56 \times 10^{26} \text{ MeV.}$$

6. *Energy released in the fission of a single uranium nucleus is 200 MeV. Calculate the number of fissions per second to produce 1 milliwatt power.*

[Ans. 3.125×10^{7}]

7. A reactor is developing nuclear energy at a rate of 32,000 kilowatts. How many atoms of U^{235} undergo fission per second? How many kg of U^{235} would be used up in 1000 hours of operation ? Assume an average energy of 200 MeV released per fission. Take Avogadro's number as 6×10^{13} and 1 MeV = 1.6 $\times$ 10^{-23} joule.

Solution: The power developed by the reactor is 32000 kilowatts, i.e. 3.2×10^7 watts. Therefore, the energy released by the reactor per second is

$$= 3.2 \times 10^7 \text{ joules} \quad [\therefore 1 \text{ watts} = 1 \text{ joule/sec}]$$

$$= \frac{3.2 \times 10^7}{1.6 \times 10^{-13}} \quad [\therefore 1.6 \times 10^{-13} \text{ joule} = 1 \text{ MeV}]$$

$$= 2.0 \times 10^{20} \text{ MeV}.$$

The energy released per fission is 200 MeV. Therefore, the number of fissions occurring in the reactor per second

$$= \frac{2.0 \times 10^{20}}{200} = 1.0 \times 10^{18}.$$

The number of atoms (or nuclei) of U^{235} consumed in 1000 hours

$$= 1.0 \times 10^{18} \times (1000 \times 3600)$$

$$= 36 \times 10^{23}.$$

Now, 1 gm-atom (*i.e.* 235 gm) of U^{235} has 6×10^{23} atoms. Therefore, the mass of U^{235} consumed in 1000 hours is

$$= \frac{36 \times 10^{23}}{6 \times 10^{23}} \times 235$$

$$= 1410 \text{ gm} = 1.41 \text{ kg}.$$

8. Calculate the approximate mass of uranium which must undergo fission to produce same energy as is produced by the combustion of 10^5 kg of coal. Heat of combustion of coal is

8000 kcal/kg ; the energy released per fission of U^{235} is 200 MeV and Avogadro's number is 6.02×10^{23} per gm-atom (1 cal = 4.2 joules).

Solution: The energy produced by 10^5 kg of coal

$$= 10^5 \times 8000 \text{ kcal}$$

$$= 8 \times 10^{11} \text{ cal}$$

$$= 8 \times 10^{11} \times 4.2 = 3.36 \times 10^{12} \text{joule}$$

$$[\because 1 \text{ cal} = 4.2 \text{ joule}$$

$$= \frac{3.36 \times 10^{12}}{1.6 \times 10^{-13}} = 2.1 \times 10^{25} \text{ MeV}$$

$$[\therefore 1 \text{ MeV} = 1.6 \times 10^{-13} \text{ joule}]$$

The energy released per fission is 200 MeV. Therefore, the number of fissions required for 2.1×10^{25} MeV energy is

$$\frac{2.1 \times 10^{25}}{200} = 1.05 \times 10^{23}.$$

Since 235 gm of U^{235} contains 6.02×10^{23} atoms, the mass of uranium containing 1.05×10^{23} atoms is

$$\frac{1.05 \times 10^{23}}{6.02 \times 10^{23}} \times 235 = 41 \text{ gm}.$$

9. *Certain stars obtain part of their energy by the fusion of three α–particles to form a $_6C^{12}$ nucleus. How much energy does each such reaction evolve ? The mass of helium atom is 4.00260 amu while the mass of an electron is 0.00055 amu. The mass of $_6C^{12}$ atom is 12.0000... amu by definition. (1 amu = 931.5 MeV)*

Solution: The mass of $_2He^4$ atom is 4.00260 amu and it has 2 electrons. Therefore, the mass of its nucleus (α-particle)

$$= 4.00260 \text{—mass of 2 electrons}$$

$$= 4.00260 - (2 \times 0.00055)$$

$$= 4.00260 - 0.00110$$

$$= 4.00150 \text{ amu.}$$

The mass of $_6C^{12}$ atom is 12.00000 amu and it has 6 electrons. Therefore, the mass of $_6C^{12}$ nucleus

$$= 12.00000 - (6 \times 0.00055)$$

$$= 12.00000 - 0.00330$$

$$= 11.99670 \text{ amu.}$$

When 3 α-particles fuse in a $_6C^{12}$ nucleus, the mass-loss is

$$(3 \times 4.00150) - 11.99670 = 0.00780 \text{ amu.}$$

The equivalent energy is

$$0.00780 \times 931.5 = 7.26 \text{ MeV.}$$

2

The Nucleus

Nuclear Structure

Nuclear Sizes and Shapes : The Rutherford α-scattering experiment established that the mass of an atom is concentrated within a small, positively-charged region at the centre which is called the 'nucleus' of the atom. Since Rutherford's time many scattering experiments, using highly energetic electrons and neutrons as the scattering particles, have been performed to determine the size of the nucleus. An electron interacts with a nucleus only through electric forces, while a neutron interacts only through nuclear forces. Thus electron scattering tells the distribution of charge in a nucleus and neutron scattering tells the distribution of nuclear mass. These experiments have shown that the volume of a nucleus is directly proportional to the number of nucleons in it, which is its mass number A.

If a nuclear radius is R, the volume is $\frac{4}{3}\pi R^3$ and so R^3 is proportional to A. This relationship is written as

$$R = R_0 A^{1/3}.$$

The value of the constant R_0 is experimentally found to be given by

$$R_0 \simeq 1.2 \times 10^{-15} \text{ m.}$$

The length 10^{-16} m is described as 1 femtometer (fm) or 1 fermi. Thus

$$1 \text{ fermi} = 10^{-15} \, m = 10^{-5} \text{ Å.}$$

Hence we can write

$$R = 1.2 \, A^{1\backslash 3} \text{ fm}$$

for nuclear radii. The radius varies from nucleus to nucleus and is of the order of 1 fm or 10^{-15} m.

While the size of the nuclei are readily determined by experiment, not too much is known about the shape of nuclei. For almost all purposes nuclei may be regarded as being 'spherical'. Certain nuclei, however, deviate from sphericity but the deviation is only 10 percent, or so.

Structure of the Nucleus—Proton-Electron Hypothesis : Before the discovery of the neutron, the nucleus of an atom was thought to be composed of protons and electrons which could account for the mass and charge of the nucleus. For example, the helium nucleus, whose mass is four times that of the proton but its charge is only + 2*e*, was thought as being composed of 4 protons and 2 electrons. This proton-electron hypothesis of the nucleus could easily account for the β–emission from the radioactive nuclei. It, however, failed because of a number of strong arguments against it :

(i) *Nuclear Size :* The radii of the nuclei are only of the order of 10^{-15} meter. The uncertainty principle demands that an electron confined to such a small region must have a kinetic energy of the order of 100 MeV. However, the electrons emitted during β–emission have energies of only 2 or 3 MeV. Hence they cannot be present within the nucleus.

The uncertainty principle applied to protons in a nucleus demand a kinetic energy only of the order of 1 MeV which is acceptable.

(ii) *Nuclear Spin :* A proton and an electron each has a spin of 1/2. Thus nuclei with an even number of protons *plus* electrons should have integral spins, while those with an odd number should have half-integral spins. This prediction is, however, not obeyed. The deuteron (nucleus of deuterium) has a mass about two times that of proton, but a charge of only + *e*. Therefore, on the proton-electron hypothesis, it must have two protons and one electron *i.e.* an odd number of particles. Its spin should therefore be half-integral $\frac{1}{2}$ or $\frac{3}{2}$ depending on the orientation of the particles). However, the observed spin of the deuteron is 1. Again, this shows that the electron as a separate particle cannot exist within the nucleus.

(iii) *Nuclear Magnetic Moment :* An electron, owing to its spin, has a magnetic moment of the order of a Bohr-magneton. Where there is an odd number of electrons, the odd electron should add this amount of magnetic moment to the nucleus. Yet the magnetic moment of nuclei are only about 1/1000 th of this. Hence the electrons cannot be the nuclear constituents.

In addition to the above theoretical considerations, experiments on electron scattering by nuclei and on nuclear disintegrations also rule out the existence of electrons in nuclei. Nevertheless, there remains the fact that many radioactive nuclei emit fast electrons (β–rays).

Proton-Neutron Hypothesis : In 1932 came the discovery of neutron, an *uncharged* particle of mass slightly more than that of a proton, and a spin of $\frac{1}{2}$. All these properties fit perfectly with the observed properties of nuclei when it is assumed that nuclei contains protons and neutrons tightly packed together (Fig.). The protons provide the (positive) charge, and the protons and neutrons together provide the mass of the nucleus. Thus the nucleus of hydrogen atom consists of 1 proton only. The nucleus of helium atom has 2 protons and 2 neutrons. In general, the nucleus of an atom of atomic number Z and mass number *A* contains Z protons and *(A–Z)* neutrons. The nuclear particles, proton and neutron, bear a common name 'nucleon'.

The hypothesis that nuclei are composed of protons and neutrons explains the existence of isotopes. The nuclei of all the isotopes of an element contain the same number of protons but have different number of neutrons. For example, ordinary hydrogen has three isotopes : hydrogen, heavy hydrogen (deuterium) and tritium. The hydrogen nucleus has only one proton, the heavy-hydrogen nucleus (called deuteron) has one proton and one neutron, while tritium nucleus (called tritron) has one proton and two neutrons.

The proton-neutron hypothesis also explains the observations regarding nuclear stability. Not all combinations of protons and neutrons form stable nuclei. There are three observations :

(i) Light nuclei contain roughly equal numbers of protons and neutrons. For heavier nuclei, the neutron number becomes progressively larger than the proton number. For the heaviest stable nuclei, the neutron number exceeds the proton number by about 50%. The excess of neutrons over protons in the heavier nuclei is necessary because protons repel one another electrostatically, and so their relative number should be less in order to secure the stability of the nucleus.

(ii) Nuclei of atoms beyond $_{83}Bi^{209}$ are unstable and undergo radioactive decay. This is so because when the proton number exceeds 83, the coulomb repulsion becomes so large that the neutrons are not able to prevent the disruption of the nucleus.

(iii) Sixty percent of the stable nuclei have even number of protons and even number of neutrons. (These are called even-even nuclei). Nearly all the others have either even number of protons and odd number of neutrons (even-odd nuclei) or vice versa (odd-even nuclei). Only five stable nuclei are odd-odd. Thus there is a preference for "even" numbers of protons or neutrons. This is in consistent with the existence of nuclear energy levels, each capable of containing two particles (neutrons or protons) of opposite spins.

Packing Fraction : The masses of the atomic isotopes are all very close to whole numbers, but are not *exactly* whole numbers. The difference between the actual atomic mass *M* of an isotope and its mass number *A* divided by the mass number is defined as the packing fraction *P* of the isotope. Thus

$$P = \frac{M - A}{A}$$

The packing fraction is positive for isotopes of very low and very high mass numbers, and negative for the rest. It is zero for $_6C^{12}$, by definition.

Mass Defect and Binding Energy : The masses of *all* stable nuclei are less than the sum of the masses of their constituent particles (protons and neutrons) in the free state. This means that when protons and neutrons combine to form a nucleus, a loss in mass results. The "missing" mass is released in the form of energy when the nucleus is formed. If Δm be the missing mass in the formation of a nucleus, the energy released ΔE will be given by the Einstein's relation :

$$\Delta E = (\Delta m)\, c^2$$

where c is the speed of light. The missing mass (Am) is called the 'mass defect', and its energy equivalent *(ΔE)* is called the 'binding energy' of the nucleus. ΔE is actually the energy that must be supplied to the nucleus to break it into its constituent particles.

The greater its binding energy, the more the energy that would be needed to break the nucleus. Thus, the binding energy is a measure of the stability of the nucleus.

Binding Energy of Deuteron : Deuteron is the simplest nucleus containing more than one nucleon. It consists of a proton and a neutron. The mass of the deuteron is less than the sum of the masses of proton and neutron. The energy equivalence of this mass difference is the binding energy of the deuteron. Then, if m_n , m_p and m_d be respectively the masses of neutron, proton and deuteron, the binding energy of the deuteron would be

$$E_B = (m_n + m_p - m_d)\, c^2. \qquad \text{... }(i)$$

In order to calculate the binding energy of the nucleus, we must use *nuclear* masses, *i.e.* neutron mass, proton mass and the mass of the nuclide.

We can write the nuclear binding energy in terms of atomic masses also. For example, the above equation will be written as

$$E_B = [m_n + m\,({}_1H^1) - m({}_1H^2)]\, c^2 \qquad \text{... }(ii)$$

where m (${}_1H^1$) stands for the mass of hydrogen *atom* and m (${}_1H^2$) for the mass of deuterium *atom.* The nuclear mass of hydrogen (the proton mass) is just the *atomic* mass of hydrogen minus the mass of one electron. The nuclear mass of deuterium (the deuteron mass) is the *atomic* mass of deuterium minus the mass of one electron. Thus electron mass automatically cencels out in calculating the nuclear binding energy from eq. (*ii*).

In general, the *nuclear* binding energy of an atom $_zx^A$ can be written in terms of atomic masses in the following way :

$$E_B = [(A - Z)m_n + Zm(_1H^1) - m(_zX^A)]\,c^2.$$

Substituting $m_n = 1.675 \times 10^{-27}$ kg, $m_p = 1.672 \times 10^{-27}$ kg,

$m_d = 3.343 \times 10^{-27}$ kg and c

$= 3.0 \times 10^8$ m/s in eq. (*i*), we get

$$E_B = (1.675 + 1.672\ 3.343) \times 10^{-27}\ \text{kg} \times (3.0 \times 10^8\ \text{m/s})^2$$

$$= 0.036 \times 10^{-11}\ \text{joule}$$

$$= \frac{0.036 \times 10^{-11}}{1.6 \times 10^{-13}} = 2.25\ \text{MeV}.$$

Thus the deuteron has a binding energy of 2.25 Me V. Using more accurate values of masses, the binding energy is about 2.23 MeV. This is confirmed by photo disintegration experiments which show that a γ-ray photon with a minimum energy of 2.23 MeV can split a deuteron into a free neutron and a free proton.

Binding Energy Curve : The nuclear binding energies range from 2.23 MeV for the heavy hydrogen to about 1800 MeV for Uranium. This shows that the binding energy increases with the complexity of the nucleus. Therefore, in order to compare the stabilities of different nuclei, we require the 'average binding energy per nucleon', which is obtained by dividing the total binding energy of a nucleus by the number of nucleons (protons+ neutrons) it contains. For example, the average binding energy per nucleon for, H^2 is

$$\frac{\text{total binding energy of } {}_1H^2}{\text{mass number (no. of nucleons)}} = \frac{2.23 \text{ MeV}}{2} = 1.115 \text{ MeV}.$$

When the average binding energy per nuclear for various nuclei is plotted against the mass number A, a curve is obtained having the appearance as shown in Fig. This is known as the 'binding energy curve. The curve rises first rapidly and then slowly until it reaches a maximum of 8.8 MeV at $A = 56$, corresponding to the iron nucleus Fe^{56}. It then drops *very slowly* to about 7.6 MeV at ,4=238 (Uranium). Evidently *nuclei of intermediate mass are the most stable,* since they have the greatest average binding energy per nucleon (in the neighbourhood of 8.5 MeV). This means that the greatest amount of energy is needed to break them in their nucleons.

The (light) nuclei with $A < 20$ are the least stable. For example, the average binding energy per nucleon for the heavy hydrogen nucleus H^2 is only about 1 MeV. The subsidiary peaks occurring at He^4, C^{12}, O^{16} (even-even nuclei) indicate that these nuclei are much more stable than their immediate neighbours.

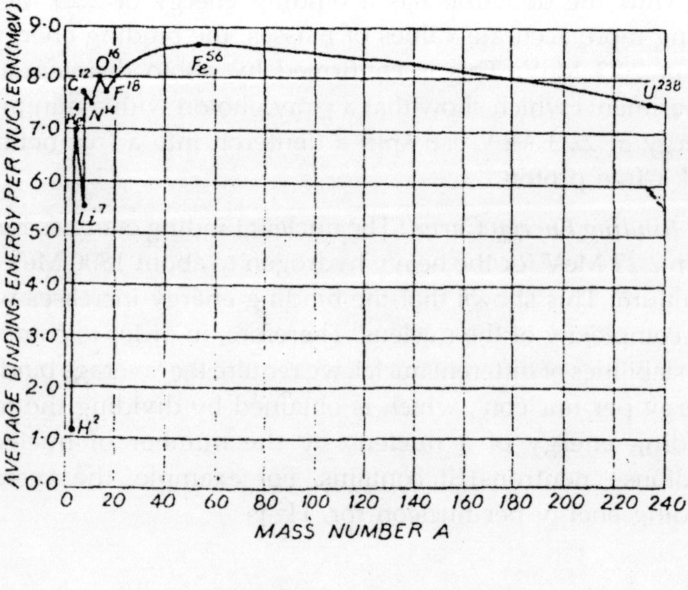

Explanation of Nuclear Fission and Nuclear Fusion : The general appearance of the binding energy versus mass number curve leads us to two remarkable conclusions which are of immense practical importance. The curve shows that very heavy and very light nuclei, which lie on either side of the broad maximum of the curve, have a lower average binding energy per nucleon. Therefore, if a very heavy nucleus (such as U^{238}) be splitted into two lighter nuclei near the top of the curve, the binding energy per nucleon will increase by about 1 MeV. This will obviously result in a general release of energy. This method of releasing nuclear energy by breaking up a heavy nucleus is known as 'nuclear fission', and is the basis of nuclear bombs and nuclear reactors.

Alternatively, if two or more very light nuclei (such as H^2) be combined into a heavier nucleus (say He^4), the binding energy per nucleon will again increase and now by a much greater amount than in the fission process. This will result in a much larger release of energy. This method of releasing nuclear energy is called 'nuclear fusion', and is the source of energy in the sun and in stars.

Nuclear Forces

The protons which repel each other electrostatically, and the neutrons which exert no electric force are held extremely close together inside a stable nucleus. Obviously, there are some forces operating within the nucleus which bind neutrons to protons (*n-p* forces), protons to protons (*p-p* forces), and neutrons to neutrons (*n-n* forces), thus holding the nucleus in one piece. Experimental evidences lead to the following properties and the nature of these nuclear forces :

(i) *Nuclear Forces are Primarily Attractive* : The overall effect of the nuclear forces is attractive, otherwise the nucleus would be disrupted under the electrostatic repulsion between the protons. (However, there is a repulsive component also in nuclear forces because an exclusively attractive force would lead to a collapse of the nucleus into a volume much smaller than the observed value).

(ii) *Nuclear Forces are Non-electric:* The nuclear forces cannot be of an electrical nature. If it were so, the protons would repel one another thus leading to nuclear disruption rather than to stability.

(iii) *Nuclear Forces are Non-gravitational :* If the gravitational forces between the nucleons are calculated, they are found to be about 10^{-40} times than the attractive forces demanded. Thus the nuclear forces cannot be gravitational in origin. In general, the nuclear forces are not central forces.

(iv) *Nuclear Forces are Extremely Strong :* The study of simplest stable nuclei, like ${}_1H^2$, shows that the forces holding the nucleons together must be very strong. In fact, they constitute by far the strongest class of forces known.

(v) *Nuclear Forces are Extremely Short-range Forces :* The nuclear forces do not obey the inverse-square law, and act only over distances as short as a small multiple of 10^{-15} meter. At distances appreciably smaller than 10^{-15} meter, the attraction of nucleons is replaced by repulsion.

(vi) *Nuclear Forces are Charge-independent :* The nuclear forces make no distinction between neutrons and protons. Experiments involving the scattering of protons and neutrons show that the (attractive) nuclear forces existing between protons and protons (*p-p* forces), between neutrons and neutrons (*n-n* forces), and between protons and neutrons (*p-n* forces) are all essentially the same in magnitude.

(vii) *Nuclear Forces are Spin-dependent :* Experimental evidences show that the nuclear force acting between two nucleons depends on the mutual orientation of the spins of the nucleons. For example, a neutron and a proton are kept together forming a deuteron only if their spins are parallel to each other.

(viii) *Nuclear Forces have the Property of Saturation* : This means that in the nucleus any one nucleon interacts with only a limited number of other nucleons nearest to it *(not* with all the other nucleons of the nucleus). This is apparent from the fact that the average binding energy per nucleon, after rising rapidly among the lightest nuclei, remains approximately constant for further increase in mass number A. This means that the total binding energy is proportional to the mass number A. When a fresh nucleon is added to the nucleus, it interacts with a *fixed* limited number of nearest nucleons, and so adds a *constant* amount to the binding energy.

***Protons in the Nucleus do not Fly Apart* :** The very Strong attractive nuclear forces (p-p, n-n and n-p) between nucleons overcome the relatively weak electrostatic repulsive forces between protons. Hence the nucleus remains intact.

The electrostatic repulsion between protons make, however, the nucleus less stable than it otherwise would be. The emission of α-particles from heavier nuclei (which have a large number of protons) indicates this fact.

***Yukawa's Meson Theory of Nuclear Forces* :** The mechanism responsible for the strongly-attractive, non-electrical and non-gravitational, short-range forces between nucleons remained a mystery for quite long. Heisenberg, in 1932, had suggested that the nuclear forces are 'exchange forces' which arise by the exchange of electrons (or positrons) between nucleons. But, calculations based on β–decay data snowed that the forces resulting from electron and positron exchange were 10^{14} times weaker than those actually required for nuclear stability.

Yukawa, in 1935, predicted the existence of a new particle, now called a pi-meson, having a rest mass greater than that of an electron but less than that of a nucleon. He proposed a theory of nuclear forces which involves these pi-mesons and is known as the 'meson theory of nuclear forces.' According to this theory, all nucleons (protons and neutrons) consist of identical cores surrounded by a 'cloud' of one or more pi-mesons. The mesons may be neutral, or may carry either a positive or a negative charge equal to the electronic charge, and are designated as π^o, π^+ and π^- respectively. The sole difference between protons and neutrons lies in the composition of their respective meson clouds. The mesons are being continually exchanged between nearby nucleons. The (nuclear) force between a neutron and a proton is the result of the exchange of charged mesons (π^+ and π^-) between them. When a π^+ meson jumps from a proton to a neutron, the proton is converted into a neutron, and the neutron is converted into a proton :

$$p - \pi^+ \rightarrow n$$

and $$p + \pi^+ \rightarrow p.$$

Conversely, when a π^- meson jumps from a neutron to a proton, the neutron is converted into a proton, and the proton is converted into a neutron :

$$n - \pi^- \rightarrow p$$

and $$p + \pi^- \rightarrow n.$$

In the same way, the forces between two protons and those between two neutrons arise by the exchange of neutral mesons (π°) between them. Thus the nucleus is an ever-changing structure. However, the numbers of protons and neutrons remain the same in a particular structure.

A question arise at this point : When mesons have a finite rest mass, why this mass is not added to the mass of protons and neutrons ? The answer lies in the uncertainty principle. The mesons are constantly running between the nucleons. That is, the nucleons are constantly emitting and absorbing mesons. As such, even in principle, it is impossible to detect any change in mass of the nuclei's.

The actual discovery of pi-mesons in cosmic radiation, after Yukawa's prediction of its existence, opened a new field of research in nuclear structure.

Estimation of π-meson Mass : Let m_π be the mass of the π-meson. Its energy equivalence is $\Delta E = m_\pi c^2$. Suppose the meson travels between nucleons at a speed v ($\approx c$). Nuclear forces have a maximum range r of about 1.7 fm ($= 1.7 \times 10^{-15}$ m). The time Δt required for the meson to travel this distance is

$$\Delta t = \frac{r}{v} \approx \frac{r}{c}$$

From the uncertainty relation, we have

$$\Delta E\ \Delta t \approx \frac{h}{2\pi}$$

or $$(m\pi c^2)\left(\frac{r}{c}\right) \approx \frac{h}{2\pi}$$

or $$m_\pi \approx \frac{h}{2\pi}\frac{1}{rc}.$$

Substituting the values, we get

$$m_\pi \approx \frac{6.63 \times 10^{-34}\ \text{j-s}}{2 \times 3.14 \times \left(1.7 \times 10^{-15}\ \text{m}\right) \times \left(3.0 \times 10^{8}\ \text{m/s}\right)}$$

$$\approx 2.1 \times 10^{-28}\ \text{kg}.$$

This is about 230 times the rest mass of the electron (9.1×10^{-31} kg).

Nuclear Models : An atomic nucleus shows a number of properties: It has (*i*) a radius proportional to $A^{1/3}$ (where A is the mass number), (*ii*) a well-defined and uniform density, (*iii*) a binding energy per nucleon which is almost constant (except for very small *A)* and (*iv*) a neutron to proton ratio which is close to unity for small *A* but increases progressively with increasing *A*. Further, it shows many types of nuclear reactions, radioactive decay and also the phenomena of nuclear fission and nuclear fusion. Many nuclear models have been suggested in an attempt to explain these properties although none of them is entirely satisfactory. Of these, two are more important : the liquid-drop model and the shell model.

Liquid-drop Model : Bohr, in 1936, pointed out that the strong, short-range, attractive forces between the nucleons are analogous to those acting between the molecules of a liquid. Therefore, *the nucleus is like a liquid-drop in which the nucleons are closely packed together, each nucleon being strongly attracted only by its immediate neighbours.* On this model, the nucleus will be a sphere with a radius proportional to $(\text{mass})^{1/3}$, and its density will be independent of its volume.

Let us now see how the liquid-drop model accounts for the observed variation of binding energy per nucleon with mass number. We know that the binding energy arises from the proton-proton, neutron-neutron, and proton-neutron nuclear attractions and is proportional to the number of nucleons *A* (the mass number). We may write it as

$$Ev = a_1 A,$$

where a_1 is a positive constant. The energy E_v is called the volume energy of the nucleus.

Now, treating the nucleus as a liquid-drop, the binding energy is decreased by a surface term, because the nucleons on the surface have fewer neighbours to attract than those in the interior. This is proportional to the surface area, so that

$$E_s = -a_2 A^{2/3}.$$

The negative energy E_s is called the surface energy of the nucleus. It is most significant for the lighter nuclei since a greater fraction of their nucleons are on the surface.

The electrostatic repulsion between each pair of protons in the nucleus also contributes toward decreasing its binding energy. The coulomb energy is proportional to the number of proton pairs Z(Z—1)/2 ! in the nucleus, and inversely proportional to the nuclear radius ($\propto A^{1/3}$), so that

$$E_c = -a_3 \frac{Z(Z-1)}{A^{1/3}}.$$

The coulomb energy is negative because it arises from an effect that opposes nuclear stability.

The total binding energy of a nucleus is the sum of its volume, surface and coulomb energies :

$$E_B = E_v + E_s + E_c$$
$$= a_1 A - a_2 A^{2/3} - a_3 Z(Z-1) A^{-1}/3.$$

In a more complete equation two more energy terms are added *to* account for the excess of neutrons over protons in the heavier nuclei and for the occurrence of *proton pairs* and neutron pairs. These energies are called 'asymmetry energy' and 'pairing energy' respectively. The asymmetry energy is negative because it reduces the binding energy of the nucleus. The pairing energy is positive for even-even nuclei (which are most stable), zero for odd-even and even-odd nuclei, and negative for odd-odd nuclei. The average binding energy *per nucleon* is

$$\frac{E_B}{A} = a_1 - a_2 A^{-1/3} - a_3 Z(Z-1) A^{-4/3}.$$

In Fig., each of the terms on the right is plotted against A, together with their sum, E_B/A. The curve representing the *total* binding energy per nucleon is fairly close to the experimental curve. Hence the analogy of a nucleus with a liquid drop has some validity at least.

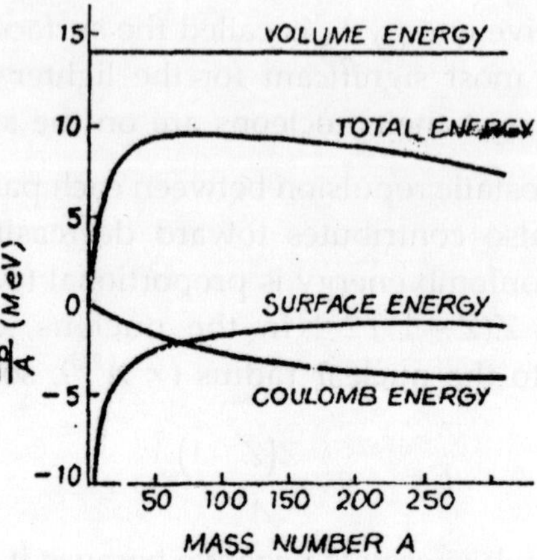

The nuclear reactions, accompanied by the emission of nucleons from the nucleus, can be interpreted on the basis of liquid-drop nuclear model. If the thermal agitation in the drop of a liquid is increased by heating, evaporation of molecules takes place. In a similar way, when an energetic particle is captured by a nucleus, a *compound* nucleus is formed which emits nucleons almost immediately. The emission of nucleons from a compound nucleus is thus analogous to the evaporation of molecules from a hot liquid drop.

Nuclear Fission and Liquid-drop Model : The liquid-drop model best explains the nuclear fission in which a heavy nucleus splits into two more or less equal parts. This occurs in the same way as a liquid-drop, if made to oscillate with sufficient energy, may become unstable and break.

The liquid-drop model fails, however, to explain nuclear reactions involving very high energies, and also the properties of the excited states of the nuclei.

Shell Model—Magic Numbers : There are certain nuclear phenomena which show the existence of excited states of nuclei and of sharply defined nuclear energy levels. This led to the shell model of the nucleus, the idea of which was borrowed from external electronic structure of the atom.

We know that the electrons in an atom are grouped in "shells", and "sub-shells", and atoms with 2, 10, 18, 36, 54, and 86 electrons have all their shells completely filled. Such atoms are unusually stable and are chemically inert. A similar situation exists with nuclei also. *Nuclei having 2, 8, 20, 28, 50, 82 and 126 nucleons of the same kind (either protons or neutrons) are more stable than nuclei of neighbouring mass numbers.* These numbers are called as 'magic numbers'. This observation is based on the following lines :

(i) Helium (Z = 2) and oxygen (Z=8) are particularly stable.

(ii) Calcium (Z = 20) and tin (Z = 50) have larger number of stable isotopes than any other element.

(iii) The end-products of the three naturally occurring radioactive series are lead isotopes (Z = 82).

(iv) The binding-energy curve has small peaks (showing extra stability) for the following nuclei :

$$_{28}Ni^{62},\ _{38}Sr^{88},\ _{50}Sn^{120},\ _{58}Ce^{140},\ _{82}Pb^{208}$$

It is seen that for these nuclei either Z (number of protons) or $A - Z$ (number of neutrons) is a magic number.

(v) The electric quadripole moments of magic number nuclei are very low (nearly zero) compared with those of other nuclei. This means that these nuclei have almost *spherical* charge distribution which is expected for the more stable nuclei.

The shell model of the nucleus is an attempt to account for the existence of magic numbers and certain other nuclear properties. According to this model, *the nucleons (protons and neutrons) in a nucleus live in a square potential well with rounded corners in which they occupy separate discrete sets of energy levels.* When a spin-orbit interaction is assumed to exist, the energy levels of either class of nucleon are arranged into different groups (shells). In this group-formation the gaps appear at just the right places, *i.e.*, after the completed groups having 2, 8, 20, 28, 50, 82, and 126 protons or neutrons.

The shell model is able to account for several nuclear phenomena in addition to magic numbers. For instance, it is observed that even-even nuclei are, in general, more stable than the odd-odd nuclei. This is obvious from the shell model. According to Pauli's principle, a single energy sub-level can have a maximum of two nucleons (one with spin up and other with spin down). Therefore, in an even-even nucleus only *completed* sub-levels are present which means greater liability. On the other hand, an odd-odd nucleus contains incomplete sub-levels for both kinds of nuclear which means lesser stability.

Further, the shell model is able to predict the total angular momenta of nuclei. Since the two protons (or neutrons) of each sub-level have *opposite* spins, they cancel each other's angular momentum. Hence, the actual total angular momentum of a nucleus depends only on the presence of an *extra* neutron or proton or both. In even-even nuclei, all protons and also all neutrons mutually cancel, so that the total angular momentum is zero for such nuclei. This has actually been observed with no exception.

The odd-odd nuclei have an extra proton and an extra neutron. Since each of these particles has an half-integral spin they give an integral total angular momentum, and this is experimentally observed. A similar reasoning predicts a half-integral total angular momentum for even-odd or odd-even nuclei, which is also experimentally confirmed.

Both the liquid-drop and shell models of the nucleus account, in their very different ways, different groups of nuclear phenomena. Neither is entirely satisfactory. Recently attempts have been made to combine the best features of each of these models and some success has been achieved. The resulting model is known as 'collective model'.

PROBLEMS

1. *Compute the approximate nuclear radius of Al^{27}.*

Solution: The nuclear radius of an atom of mass number A is given by

$$R = R_0 A^{1/3}$$

The proportionality constant R_0 is roughly 1.2 fermi. Thus

$$R = 1.2\ A^{1/3} \text{fm}.$$

For aluminium, $A = 27$.

$$\therefore \qquad R = 1.2\ (27)^{1/3}\ \text{fm}$$

$$= 3.6\ \text{fm} = 3.6 \times 10^{-15}\ \text{m}.$$

2. *A nucleus with A = 235 splits into two nuclei whose mass numbers are in the ratio 2:1. Find the radii of the nuclei. (Take R_0 = 1.4 fermi).*

Solution: The mass numbers of the resulting nuclei are 157 and 78.

The radius of a nucleus of mass number A is given by

$$R = R_0\ A^{1/3}$$

$$= (1.4\ \text{fm})\ A^{1/3}. \qquad [\text{Given} : R_0 = 1.4\ \text{fm}]$$

For $A = 57$, we have

$$R = (1.4\ \text{fm})\ (157)^{1/3} = 7.5\ \text{fm}.$$

For $A = 78$, we have

$$R = (1.4\ \text{fm})\ (78)^{1/3} = 6.0\ \text{fm}.$$

3. *What is the mass number A of a nucleus whose radius R is 2.5 fm ?*

Solution: $\qquad R = R_0\ A^{1/3},$

where R_0 is roughly 1.2 fm (1 fm = 10^{-15} m). Here R = 2.5 fm.

$$\therefore \qquad 1.2\ A^{1/3} = 2.5$$

$$\therefore \quad 1.2\, A^{1/3} = 2.5$$

$$\text{or} \quad A^{1/3} = \frac{2.5}{1.2} = 2.08,$$

$$\therefore \quad A = (2.08)^3 = 9.$$

4. *Estimate the density of nuclear matter in kg/m^3, given that the nuclear radius R≈l.2 A$^{1/3}$ fm. Take mass of the nucleus A amu where A is mass number and 1 amu = l.66 × 10^{-27} kg. (1 fm = 10^{-15} meter.)*

Solution: The mass of the nucleus is

$$M = A \text{ amu} = A \times 1.66 \times 10^{-27} \text{ kg.}$$

Treating the nucleus as sphere, its volume is

$$V = \frac{4}{3}\pi R^3 \approx \frac{4}{3} \times 3.14 \times (1.2\, A^{1/3} \times 10^{-15} \text{ meter})^3$$

$$\frac{4}{3} \times 3.14 \times (1.2)^3\, A \times 10^{-45} \text{ meter}^3$$

$$- 7.23 \times 10^{-45}\, A \text{ meter}^3.$$

Therefore, the density of nuclear matter is given by

$$\rho = \frac{M}{V} \approx \frac{A \times 1.66 \times 10^{-27} \text{ kg}}{7.23 \times 10^{-45} \text{ A meter}^3}$$

$$\approx 2.3 \times 10^{17} \text{ kg/m}^3.$$

5. *Calculate the binding energy per nucleon in $_6C^{12}$. Masses of proton, neutron and electron are 1.00728, 1.00867 and 0.00055 amu respectively. The mass of $_6C^{12}$ atom is 12.00000 amu.*

Solution: The nucleus of $_6C^{12}$ atom contains 6 protons and 6 neutrons. Now,

mass of 6 protons = 6 ×1.00728 = 6.04368 amu

mass of 6 neutrons = 6 × 1.00867 = 6.05202 amu

total = 12.09570 amu.

This is the mass of the constituent particles of the carbon nucleus.

The carbon atom has 6 electrons. Now,

$$\text{mass of } {}_6C^{12} \textit{ atom} = 12.00000 \text{ amu}$$

$$\text{mass of 6 electrons} = 6 \times 0.00055 = 0.00330 \text{ amu}$$

$$\text{differences} = 11.99670 \text{ amu.}$$

This is the mass of the ${}_6C^{12}$. *nucleus.* Therefore, the mass defect is

$$\Delta m = \text{mass of nucleons—mass of nucleus}$$

$$= 12.09570 - 11.99670 = 0.09900 \text{ amu}$$

We know that 1 amu is equivalent to 931.5 MeV of energy. Therefore, the energy equivalent of Δm is

$$\Delta E = 0.09900 \times 931.5 = 92.2 \text{ MeV.}$$

This is the binding energy of carbon nucleus which has 12 nucleons. Therefore binding energy per nucleon is

$$\frac{92.2}{12} = 7.68 \text{ MeV.}$$

6. ***Taking the result of the last problem, calculate the binding energy of the least strongly bound neutron in the nucleus of ${}_6C^{13}$ isotope. The mass of the isotope is 13.00335 amu.***

[Ans. 4.95 MeV]

7. ***The atomic mass of ${}_6C^{12}$ is 12.000000 amu and that of ${}_6C^{13}$ is 13.003354 amu. Find the energy required to remove a neutron from ${}_6C^{13}$ in MeV. The mass of neutron is 1.008665 amu.***

Solution: The *nuclear* equation is

$${}_6C^{13} \rightarrow {}_0C^{12} + {}_0n^1$$

The mass of $_6C^{12} + {}_0n^1$ is 12.000000 + 1.008665 = 13.008665 amu.

$$\therefore \quad \text{mass defect} = 13.008665 - 13.003354$$
$$= 0.005311 \text{ amu.}$$

Its energy equivalent is

$$\Delta E = 0.005311 \times 931.5 = 4.95 \text{ MeV.}$$

This is the required energy.

8. ***Compute the binding energy of the "last" proton in a $_6C^{12}$ nucleus if the mass of this nucleus is 11.99671 amu and the mass of the $_5B^{11}$ nucleus is 11.00657 amu. The mass of proton is 1.00728 amu.***

Solution: The *nuclear* equation is

$$_5B^{11} + {}_1p^1 \rightarrow {}_6C^{12}.$$
$$\text{(proton)}$$

The mass of $_5B^{11} + {}_1P^1$ is 11.00657 + 1.00728 = 12.01385 amu, while that of the product nucleus $_6C^{12}$ is 11.99671 amu. Therefore, mass loss is

$$\Delta m = 12.01385 - 11.99671 = 0.01714 \text{ amu.}$$

Its energy equivalent is

$$\Delta E = 0.01714 \times 931.5 = 16 \text{ MeV.}$$

This is the binding energy of the last proton.

9. ***Calculate the binding energy of α-particle in MeV. The masses of the proton, neutron and the α-particle (helium nucleus) are 1.00728,1.00867 and 4.00151 amu respectively.***

Solution: The α-particle ($_2He^4$) is composed of two protons and two neutrons. The sum of the masses of its constituent particles is as follows :

$$\text{mass of 2 protons} = 2 \times 1.00728 = 2.01456 \text{ amu}$$
$$\text{mass of 2 neutrons} = 2 \times 1.00867 = 2.01734 \text{ amu}$$
$$\text{total} = 4.03190 \text{ amu.}$$

Subtracting from this sum, the mass of the α-particle (He nucleus), we obtain the mass-defect Δm, that is,

$$\Delta m = 4.03190 - 4.00151 = 0.03039 \text{ amu.}$$

1 amu is equivalent to 931.5 M.eV of energy. Therefore the energy equivalent of the above mass defect is

$$\Delta E = 0.03039 \times 931.5 = 28.3 \text{ MeV.}$$

This is the energy that would be liberated if one were to make an α-particle out of protons and neutrons. Conversely, this is also the energy that would have to be supplied to an α-particle to break it into protons and neutrons, and consequently it is the binding energy of the α-particle. This *large* amount of energy explains why the α-particle is a highly stable particle.

The α-particle has 4 nucleons. Hence its binding energy per nucleon is $\frac{28.3}{4} = 7.07$ MeV.

10. ***How much energy must a photon have if it is to split an α-particle ($_2He^4$ nucleus) into a triton ($_1H^3$ nucleus) and a proton ($_1H^1$ nucleus) ? The masses of the α-particle, triton and proton are 4.00151, 3.01550 and 1.00728 amu respectively and 1 amu is 931.5 MeV.***

Solution: The energy required for the photon is the binding energy of the α-particle formed by the fusion of triton and proton.

The mass of triton *plus* proton is 3.01550 + 1.00728 = 4.02278 amu, while the mass of the resulting α-particle is 4.00151 amu. Thus, the missing mass is

$$\Delta m = 4.02278 - 4.00151 = 0.02127 \text{ amu.}$$

Its energy equivalent is

$$\Delta E = 0.02127 \times 931.5 = 19.8 \text{ MeV.}$$

This is the binding energy, and hence the required energy for the splitting of the α-particle.

11. ***Calculate the binding energy of a deuteron (nucleus of $_1H^2$) from the following data :***

Mass of deuterium atom $_1H^2$ = 2.014102 u, mass of hydrogen atom $_1H^1$ =1.007825 u ana mass of neutron = 1.008665 u. (l u= 1.66 × 10^{-27} kg.)

Or

Calculate the minimum energy of gamma-rays necessary to disintegrate a deuteron into a proton and a neutron.

Solution: The deuterium atom $_1H^2$contains one proton and one neutron in its nucleus (called as deuteron) and one extra-nuclear electron. Let us calculate the mass defect of deuteron. We have

$$\text{mass of } {}_1\text{H}^1 \text{ atom (1 proton and 1 electron)} = 1.007825 \text{ u}$$

$$\text{mass of 1 neutron} = 1.008665 \text{ u}$$

$$\text{total} = 2.016490 \text{ u.}$$

This is the mass of all the constituent particles of the atom $_1\text{H}^2$. The mass of the atom $_1\text{H}^2$ is 2.014102 u. Therefore, the mass defect of its *nucleus* is

$$\Delta m = 2.016490 - 2.014102 = 0.002388 \text{ u.}$$

Now, $$1 \text{ u} = 1.66 \times 10^{-27} \text{ kg.}$$

$$\therefore \quad \Delta m = 0.002388 \times (1.66 \times 10^{-27})$$

$$= 3.964 \times 10^{-30} \text{ kg.}$$

Its energy equivalent is

$$\Delta E = (Am)\ c^2$$

$$= (3.964 \times 10^{-30} \text{ kg}) \times (3.0 \times 10^8 \text{ m/s})^2$$

$$= 3.567 \times 10^{-13} \text{ joule.}$$

Now, $$1 \text{ MeV} = 1.6 \times 10^{-13} \text{ joule.}$$

$$\therefore \quad \Delta E = \frac{3.567 \times 10^{-13}}{1.6 \times 10^{-13}} = 2.23 \text{ Mev.}$$

This is the (nuclear) binding energy of the deuteron.

When a deuteron is formed from a free proton and a free neutron, then 2.23 MeV of energy is liberated. Conversely, a gamma-ray photon must have an energy of at least 2.23 MeV to break a deuteron into a proton and a neutron.

12. ***Calculate the average binding energy per nucleon for a U^{238} nucleus in MeV. The mass of U^{238} atom is 238.05078 amu and it is composed of 92 protons and 146 neutrons. The masses of proton, neutron and electron are 100728, 1.00867 and 0.00055 amu respectively.***

Solution: Let us consider U^{238} atom. Its nucleus contains 92 protons and 146 neutrons. The sum of the masses of the constituent nucleons is as follows :

$$\text{mass of 92 protons} = 92 \times 1.00728 = 92.66976 \text{ amu}$$

$$\text{mass of 146 neutrons} = 146 \times 1.00867 = 147.26582 \text{ amu}$$

$$\text{total} = 239.93558 \text{ amu.}$$

The mass of the U^{238} atom is 238.05078 amu. It contains 92 electrons. Therefore, the mass of U^{238} nucleus is

$$238.05078 - (92 \times 0.00055) = 238.00018 \text{ amu.}$$

The mass defect is

$$\Delta m = 239.93558 - 238.00018$$

$$= 1.93540 \text{ amu.}$$

Its energy equivalent, that is, the binding energy of U^{238} nucleus is

$$\Delta E = 1.93540 \times 931.5$$

$$[\because 1 \text{ amu} = 931.5 \text{ MeV}]$$

$$= 1803 \text{ MeV.}$$

The total number of nucleons is 238. Therefore binding energy per nuclear is

$$\frac{1803}{238} = 7.57 \text{ MeV/nucleon.}$$

13. ***Find the total energy liberated in the reaction*** $_3Li^7$ ***(P, α) 2He*** $_e^4$ ***if the binding energy per nucleon for*** Li^7 ***and*** He^4 ***are 5.61 MeV and 7.06 MeV respectively.***

Solution: The reaction is

$$_3Li^7 + {_1H^1} \rightarrow {_2He^4} + {_2He^4}.$$

The binding energy of $_3Li^7$ = 5.61 × 7 = 39.27 MeV and that of $_2He^4$ = 7.06 × 4 = 28.24 MeV. There is *no* binding energy in the nucleus of $_1H^1$ (it is simply a proton).

Thus, the total binding energy of the reactants is 39.27 MeV and that of the products is 2 × 28.24 = 56.48 MeV. The increase in binding energy is

$$56.48 - 39.27 = 17.21 \text{ MeV}.$$

This is the total energy liberated.

14. ***The binding energy of*** $_{17}C1^{35}$ ***nucleus is 298 MeV. Find its atomic mass. The mass of hydrogen atom (***H^1***) is 1.007825 amu and that of a neutron is 1.008665 amu. Given : 1 amu = 931.5 MeV.***

Solution: The $_{17}Cl^{35}$ atom has 17 protons and 18 neutrons in its nucleus, and 17 extra nuclear electrons. Its mass would be the sum of the masses of the constituent particles *minus* the mass equivalent of the binding energy of its nucleus. Now,

mass of 17 $_1H^1$ atoms *i.e.* 17 protons and 17 electrons

$$= 17 \times 1.007825 = 17.133025 \text{ amu}$$

mass of 18 neutrons = 18 × 1.008665 = 18.155970 amu

$$\text{total} = 35.288995 \text{ amu.}$$

The mass equivalent of the binding energy of $_{17}C1^{35}$ nucleus is

$$\Delta m = \frac{298}{931.5} \; 0.319914 \text{ amu.}$$

∴ atomic mass of $_{17}C1^{35}$

$$= 35.288995 - 0.319914 = 34.96908 \text{ amu.}$$

3

Nucleus Hypothesis

Nuclear Reactions

A nuclear reaction is a strong interaction of an atomic *nucleus* with an elementary particle, which results in a new nucleus and one or more new particles. The reaction occurs when the interacting nucleus and the particle approach each other up to distances of the order of 10^{-15} meter.

An equation representing a nuclear reaction can be written as

$$X + a \rightarrow Y + b,$$

where X is the initial nucleus, a the initial particle, Y the final nucleus and b the final particle. It can be written in the following abbreviated form :

$$X\ (a,\ b)\ Y$$

The particles participating in the reaction are indicated in parentheses, first the initial particle, then the final one.

The particles a and b may be alpha particle ($_2He^4$), proton

($_1H^1$ or p), deuteron ($_1H^2$ or d), neutron (n), gamma-ray photon (γ), etc.

Nuclear reactions are usually attended by the liberation or absorption of energy, which is called the 'reaction energy' or 'disintegration energy'.

The first nuclear reaction discovered by Rutherford is

$$_7N^{14} + {}_2He^4 \rightarrow {}_8O^{17} + {}_1H^1$$

(α-particle) (proton)

or $_7N^{14}$ (α, p) $_8O^{17}$.

In this reaction a nitrogen nucleus hit by an α-particle is converted into an oxygen nucleus and emits a proton. The reaction is called an alpha-proton (α, p) reaction.

Like a chemical reaction, a nuclear reaction must be balanced; the number of protons and neutrons must be the same on both sides of the equation.

Compound Nucleus Hypothesis

Bohr, in 1936, proposed the compound nucleus hypothesis for nuclear reactions. According to it, many nuclear reactions proceed in two separate stages :

(i) The incident particle strikes the initial (or target) nucleus and the two first combine to form a new nucleus, called a *compound nucleus.* The atomic and mass numbers of the compound nucleus are respectively the sum of the atomic numbers and the sum of the mass numbers of the incident particle and the target nucleus. The energy brought by the incident particle is quickly shared by the nucleons of the compound nucleus which is therefore invariably in an excited state.

(ii) The (excited) compound nucleus immediately breaks up by ejecting a particle or a γ–photon, leaving the final (or product) nucleus.

Thus the Rutherford's reaction can be written to show the intermediate stage of the compound nucleus formation as

$$_7N^{14} + {_2He^4} \rightarrow ({_9F^{18}}) \rightarrow {_8O^{17}} + {_1H^1}.$$

The compound nucleus has an average life-time of the order of 10^{-14} sec which is very long compared with the time ($\sim 10^{-21}$ sec) required for a particle with an energy of several MeV to pass through a nucleus. The compound nucleus therefore "forgets" how it was formed and its breaking is independent of the mode of its formation. In fact, the same compound nucleus can be formed by *different* pairs of incident particle and target nucleus, and can break in different ways, giving different products. The idea of compound nucleus has been very fruitful in correlating many nuclear phenomena.

Stripping Reaction and Pick-up Reaction : A different type of nuclear reaction is sometimes observed when a deuteron strikes a nucleus. Because of the finite distance between the proton and the neutron in the deuteron, one of these particles reaches the nuclear surface earlier and is captured by the nucleus, whereas the other particle flies past the nucleus. This is called 'stripping reaction' because a nucleon is stripped out of the deuteron by the target nucleus.

The reverse of a stripping reaction is a 'pick-up reaction'. A nucleon (*n or p*) striking a nucleus picks-up from it a nucleon (*p* or *n*) and forms a deuteron.

Conservation Laws in Nuclear Reactions : The nuclear reactions are governed by certain conservation laws which limit the possible outcomes of the reaction. Some of these laws are as follows :

1. *Conservation of Total Relativistic Energy :* Mass and kinetic energy are *not* separately conserved in nuclear reactions. Instead, the total relativistic energy (kinetic energy + rest mass energy) is conserved.

Let us consider the general case in which a bombarding particle *a* interacts with a target nucleus *X* to give a product nucleus *Y* and a product particle *b* ; that is

$$X + a \rightarrow Y + b.$$

According to the conservation law, the kinetic energy plus the rest energy on one side of the reaction equation must equal the corresponding sum on the other side ; that is

$$m_x c^2 + (K_a + m_a c^2) = (K_y + m_y c^2) + (K_b + m_b c^2).$$

Here $K_x = 0$ since X is stationary in the laboratory frame.

2. *Conservation of Linear Momentum* : The resultant linear momentum of the system (*X* + *a*) must be equal in magnitude and direction to that of the system (*Y* + *b*). It helps in determining the energy balance of the nuclear reaction.
3. *Conservation of Angular Momentum:* The sum of the spin angular momentum and the motional angular momentum of the system (*X+a*) should be equal to the corresponding sum of the system (*Y* + *b*). The angular momentum affects the rate, or cross-section, of the nuclear reaction.
4. *Conservation of Charge* : The total net electric charge before and after the nuclear reaction should be the same.
5. *Conservation of the number of Nucleons (Mass Number):* In all nuclear reactions the number of nucleons (or the mass number) remains conserved, *i.e.*, the total number of nucleons present before the reaction equals the total number present after.
6. *Conservation of Parity* : Parity is also conserved in a nuclear reaction.

Nuclear Reactions induced by α-particles : The (positively-charged) α-particles can readily penetrate light (low Z) nuclei which offer less coulombic repulsion. Besides the Rutherford's first reaction, some other reactions induced in *light* nuclei by α-particles with energies upto about 7-8 MeV emitted from natural substances are as follows :

$$\left.\begin{array}{l} {}_5B10 + {}_2He^4 \rightarrow ({}_7N^{14}) \rightarrow .6C^{13} + {}_1H^1 \\ {}_{12}Al^{27} + {}_2He^4 \rightarrow ({}_{15}P^{31}) \rightarrow {}_{14}Si^{30} + {}_1H^1 \end{array}\right\} \quad (\alpha, p) \text{ reactions}$$

$$\left.\begin{array}{l} {}_4Be^9 + {}_2He^4 \rightarrow ({}_6C^{13}) \rightarrow {}_6C^{12} + {}_0n^1 \\ {}_5B^{11} + {}_2He^4 \rightarrow ({}_7N^{15}) \rightarrow {}_7N^{14} + {}_0n^1 \\ {}_7N^{14} + {}_2He^4 \rightarrow +({}_9F^{18}) \rightarrow {}_9F^{17} + {}_0n^1 \\ {}_{13}Al^{27} + {}_2He^4 \rightarrow ({}_{15}P^{31}) \rightarrow {}_{15}P^{30} + {}_0n^1 . \end{array}\right\} \quad (\alpha, n) \text{ reactions}$$

As we are seeing, a given incident particle and a given target nucleus can lead to different reaction products depending on the energy of the incident particle.

α-particles directed towards *heavier* nuclei experience strong coulombic repulsion and are scattered rather than reacting with the nuclei.

Reactions induced by Protons : Protons (and deuterons) with a single positive charge would experience smaller coulombic repulsion than α-particles in the field of the nucleus, Hence these particles, when accelerated to high energies by means of particle-accelerators can induce reactions in many nuclei. Cockcroft and Walton were the first to produce such an artificially induced reaction in 1932. They disintegrated lithium-7 by protons accelerated, to about 0.3 MeV which resulted in α-particle production :

$$_3Li^7 + {}_1H^1 \rightarrow ({}_4Be^8) \rightarrow {}_2He^4 + {}_2He^4 \qquad (p, 2\alpha)$$

This reaction had provided one of the earliest verifications of the Einstein's mass-energy formula. The other proton-induced reactions which give rise to α-particles, neutrons, and γ–rays are as follows:

$$\left.\begin{array}{l} {}_4Be^9 + {}_1He^1 \rightarrow ({}_5B^{10}) \rightarrow {}_3Li^6 + {}_2He^4 \\ {}_9F^{19} + {}_1He^1 \rightarrow (10Ne^{20}) \rightarrow {}_8O^{16} + {}_2H^4 \\ {}_{13}Al^{27} + {}_1H^1 \rightarrow ({}_{14}Si^{28}) \rightarrow {}_{12}Mg^{24} + {}_2He^4 \end{array}\right\} \quad (p, \alpha)$$

$${}_5B^{11} + {}_1H^1 \rightarrow ({}_6C^{12}) \rightarrow {}_2He^4 + {}_2He^4 + {}_2He^4 \quad (p, 3\alpha)$$

$${}_8O^{18} + {}_1H^1 \rightarrow ({}_9F^{19}) \rightarrow {}_9F^{18} + {}_0n^1 \quad (p, n)$$

$${}_3Li^7 + {}_1H^1 \rightarrow ({}_4Be^8) \rightarrow {}_4Be^8 + \gamma \quad (p, \gamma)$$

In the last reaction the bombarding proton is captured by the target nucleus forming an excited compound nucleus which is simply de-excited by emitting γ–ray photon. Such a reaction is called 'radiative capture'.

Reactions induced by Denterons : Many nuclear reactions have been observed with deuterons accelerated to energies over a wide range upto several hundred MeV, producing α-particles, protons and neutrons.

$${}_2Li^6 + {}_1H^2 \rightarrow ({}_4Be^8) \rightarrow {}_2He^4 + {}_2He^4 \quad (d, 2\alpha)$$

$${}_8O^{16} + {}_1H^2 \rightarrow ({}_9F^{18}) \rightarrow {}_7N^{14} + {}_2He^4 \quad (d, \alpha)$$

$${}_6C^{12} + {}_1H^2 \rightarrow ({}_7N^{14}) \rightarrow {}_6C^{13} + {}_1H^1 \quad (d, p)$$

$${}_6C^{12} + {}_1H^2 \rightarrow ({}_7N^{14}) \rightarrow {}_7n^{13} + {}_0n^1 \quad (d, n)$$

Reactions induced by Neutrons : Neutrons, being uncharged, are much more likely to penetrate nuclei than are protons, deuterons or α-particles. Therefore fast as well as slow neutrons have been found extremely effective in producing nuclear disintegrations.

The reaction between a neutron and a nucleus gives rise in most cases to an α-particle, a proton, two neutrons, or a γ–ray photon (radiative capture). The actual type of reaction depends upon the energy of the bombarding neutron and the nature of the target nucleus. Among the most interesting neutron-induced reactions are the following:

$$\left.\begin{aligned} {}_3\mathrm{Li}^6 + {}_0\mathrm{n}^1 &\rightarrow ({}_3\mathrm{Li}^7) \rightarrow {}_1\mathrm{H}^3 + {}_2\mathrm{He}^4 \\ {}_5\mathrm{B}^{10} + {}_0\mathrm{n}^1 &\rightarrow ({}_5\mathrm{B}^{11}) \rightarrow {}_3\mathrm{Li}^7 + {}_2\mathrm{He}^4 \\ {}_{13}\mathrm{Al}^{27} + {}_0\mathrm{n}^1 &\rightarrow ({}_{13}\mathrm{Al}^{28}) \rightarrow {}_{11}\mathrm{Na}^{24} + {}_2\mathrm{He}^4 \end{aligned}\right\} \quad (n, \alpha)$$

$$\left.\begin{aligned} {}_7\mathrm{N}^{14} + {}_0\mathrm{n}^1 &\rightarrow ({}_7\mathrm{N}^{15}) \rightarrow {}_6\mathrm{C}^{14} + {}_1\mathrm{He}^1 \\ {}_{13}\mathrm{Al}^{27} + {}_0\mathrm{n}^1 &\rightarrow ({}_{13}\mathrm{Al}^{28}) \rightarrow {}_{12}\mathrm{Mg}^{27} + {}_1\mathrm{H}^1 \end{aligned}\right\} \quad (n, p)$$

$${}_{13}\mathrm{Al}^{27} + {}_0\mathrm{n}^1 \rightarrow ({}_{13}\mathrm{Al}^{28}) \rightarrow {}_{13}\mathrm{Al}^{26} + {}_0\mathrm{n}^1 + {}_0\mathrm{n}^1 \qquad (n, p)$$

$$\left.\begin{aligned} {}_1\mathrm{H}^1 + {}_0\mathrm{n}^1 &\rightarrow ({}_1\mathrm{H}^2) \rightarrow {}_1\mathrm{H}^2 + \gamma \\ {}_{13}\mathrm{Al}^{27} + {}_0\mathrm{n}^1 &\rightarrow ({}_{13}\mathrm{Al}^{28}) \rightarrow {}_{13}\mathrm{Al}^{28} + \gamma \\ {}_{92}\mathrm{U}^{238} + {}_0\mathrm{n}^1 &\rightarrow ({}_{92}\mathrm{U}^{239}) \rightarrow {}_{92}\mathrm{U}^{239} + \gamma \end{aligned}\right\} \quad \text{radiative } (n, \gamma) \text{ capture}$$

The *slow* neutrons, almost in all cases, induce (n, γ) reactions (radiative capture) and the product nuclei are often radioactive. These reactions are important source of artificial radio-isotopes.

Reactions induced by Photons (Photo disintegration) : Nuclei can also be disintegrated by bombardment with γ–ray photons, provided that the energy of the photons exceeds a threshold value depending on the target nucleus. For example, deuteron can be disintegrated by bombarding with 2.62-MeV γ–rays emitted from ThC. The reaction is

$${}_1\mathrm{H}^2 + \gamma \rightarrow ({}_1\mathrm{H}^2) \rightarrow {}_1\mathrm{H}^1 + {}_0\mathrm{n}^1$$

for which the threshold energy is 2.2 MeV. This is the binding energy of the proton and the neutron in the deuteron nucleus. Reactions like (γ, 2n), (γ, αn), (γ, p) have been observed by using high-energy γ–ray photons derived from a betatron.

Q-value of Nuclear Reaction : In a nuclear reaction not only the charge numbers and mass numbers balance, but also the total relativistic energy, (= kinetic energy + rest mass energy) remains conserved. Let us consider a general nuclear reaction

$$X + a = Y + b, \qquad \text{... } (i)$$

where X is the target nucleus, a is the bombarding particle, Y is the product nucleus, and b is the product particle.

Let mx, m_a, m_y and m_b be the *rest* masses of X, a, Y and b respectively; and K_a, K_y and K_b the kinetic energies of a, Y and b respectively. The target nucleus X is initially at rest so that it has no kinetic energy ($K_x = 0$). By the conservation of *total* relativistic energy in reaction (*i*), we have

$$mx\, c^2 + (K_a + m_a\, c^2) = (K_y + my\, c^2) + (K_b + M_b\, c^2). \quad \text{... } (ii)$$

Because there can be an exchange of energy between kinetic energy and rest mass energy, the final kinetic energy $K_y + K_b$ may be greater, or less, than the initial kinetic energy K_a. The difference is called the 'Q-value' or 'energy balance' of the reaction, That

$$Q = K_y + K_b - K_a. \quad \text{... } (iii)$$

Substituting the value of $K_y + K_b - K_a$ from eq. (*ii*), we obtain

$$Q = (m_X + m_a - m_y - m_b)\, c^2 \quad \text{... } (iv)$$

Thus the Q-value may be determined by measuring the energies or the masses.

The m's in eq. (iv) refer to the rest masses of the *nuclei* X *and* Y and to the rest masses of the completely ionised *nuclear* particles a and b. However, in view of the conservation of charge in nuclear reaction, we can incorporate electron masses in m's (ignoring electron binding energy) and hence can treat them as *atomic* masses.

Reactions for which Q is positive, the rest mass energy of product particles Y and b is less than that of the initial particles X and a. The rest energy surplus of X and a is converted into the kinetic energy of the product particles which exceeds the kinetic energy of the initial particles. Such reactions are called 'exothermic' or 'exoergic'.

If Q is negative, that is, the rest energy of the product particles exceeds that of the initial particles, the reaction is 'endothermic' or 'endoergic'. In this case the surplus rest energy of the product particles results in a loss of kinetic energy.

Atomic Mass Unit (amu)

Atomic mass unit (abbreviated 'u') is the unit in which atomic masses are expressed. "1 u is defined as 1/12 th the mass of an atom of $_6C^{12}$ which is the most abundant isotope of carbon". Thus, by definition, the mass of $_6C^{12}$ *atom* is exactly 12.00000.......*u*. The masses of other atoms are determined by measuring the position of the line they produce in a mass spectrograph relative to that of $_6C^{12}$.

Atomic masses refer to the masses of neutral atoms, not of stripped nuclei. Thus the masses of the orbital electrons and the mass equivalent of their binding energies. (which are very small) are included in the figure of atomic masses.

1 gm-atom of $_6C^{12}$ has N (Avogadro's number) atoms and a mass of 12 gm. Therefore, the mass of 1 atom of $_6C^{12}$ is $\frac{12}{N}$ gm.

Thus, by definition.

$$1 \text{ u} = \frac{1}{12} \text{ mass of 1 atom of } {}_6C^{12}$$

$$= \frac{1}{12}\left(\frac{12}{N} \text{ gm}\right) = \frac{1}{N} \text{ gm.}$$

But $$N = 6.023 \times 10^{23}.$$

$\therefore$ $$1 \text{ u} = \frac{1}{6.023 \times 10^{23}} = 1.66 \times 10^{-24} \text{ gm}$$

or $$\boxed{1\,u = 1.66 \times 10^{-27} \text{ kg.}}$$

Let us now calculate the energy equivalence of 1 u. We know that 1 u = 1.66×10^{-27} kg and $E = mc^2$. Thus, for 1 u, we have

$$E = 1.66 \times 10^{-27} \text{ kg} \times (3.0 \times 10^8 \text{ m/s})^2$$
$$= 1.49 \times 10^{-10} \text{ joule.}$$

But 1 $eV = 1.6 \times 10^{-19}$ joule.

$$\therefore \quad E = \frac{1.49 \times 10^{-10} \text{ joule}}{1.6 \times 10^{-19} \text{ joule / eV}}$$
$$= 0.931 \times 10^9 \text{ eV} = 931 \text{ MeV.}$$

Using more accurate values of the constants involved yields 931.5 MeV. Thus

$$\boxed{\begin{array}{l} 1\,u \equiv 931.5 \text{ MeV} \\ 1\text{ u} \times c^2 = 931.5 \text{ MeV} \end{array}}$$

PROBLEMS

1. *Complete the following nuclear reactions :*

(i) $_{17}Cl^{35} + ? \rightarrow {}_{16}S^{32} + {}_2He^4$

(ii) $_5B^{10} + ? \rightarrow {}_3Li^7 + {}_2He^4$

(iii) $_3Li^6 + ? \rightarrow {}_4Be^7 + {}_0n^1$

(iv) $_{13}Al^{27} + {}_0n^1 \rightarrow {}_{12}Mg^{27} + ?$

(v) $_4Be^9 + {}_2He^4 \rightarrow ? + {}_0n^1$

(vi) $_3Li^7$ + *proton* → *α-particle* + ?

(vii) *deuteron+deuteron* → *proton* + ?

[**Ans.** (*i*) $_1H^1$, (*ii*) $_0n^1$, (*iii*) $_1H^2$, (*iv*) $_1H^1$, (*v*) $_6C^{12}$, (*vi*) α-particle $_2He^4$, (*vii*) tritron $_1H^3$]

2. *Complete the following disintegration reactions by substituting the proper nuclide or particle for the question mark in each case :*

(i) $_{14}Si^{28}$ (*n,p*) ?

(ii) $_3Li^7$ (*p, γ*) ?

(iii) $_{13}Al^{27}$ ***(p, γ) ?***

(iv) $_{11}Na^{23}$ ***(p ; ?)*** $_{10}Ne^{20}$

(v) $_{13}A1^{27}$ ***(?, α)*** $_{11}Na^{24}$

(vi) $_{5}B^{10}$ ***(? , a)*** $_{3}Li^{7}$

(vii) $_{11}Na^{23}$ ***(d , ?)*** $_{12}Mg^{24}$

[Ans. (*i*) $_{13}Al^{28}$, (*ii*) $_{4}Be^{8}$, (*iii*) $_{14}Si^{30}$, (*iv*) $_{2}He^{4}$, (*v*) $_{0}n^{1}$ (vi, $_{0}n^{1}$, (*vii*) $_{0}n^{1}$.]

3. *Complete the following reactions :*

(i) H^{1} ***(n , γ) ?, (ii)*** Li^{7} ***(p , ?)*** Be^{7}***, (iii)*** N^{14} ***(? , p)*** O^{17}***,***

(iv) ? (n , p) Na^{24}.

[Ans. $_{1}H^{2}$, $_{0}n^{1}$, $_{2}He^{4}$, $_{12}Mg^{24}$].

4. *Calculate the energy released (Q-value) in the following reaction :*

$$_{3}Li^{6} + {}_{1}H^{2} \rightarrow {}_{2}He^{4} + {}_{2}He^{4} + Q.$$

The atomic masses are : $_{1}H^{2} = 2.014102\ u$, $_{3}Li^{6} = 6.015123\ u$, $_{2}He^{4} = 4002603\ u$. ***Given*** $lu \times c^{2} = 931.5$ ***MeV.***

Solution: The *Q*-value is initial masses minus final masses, multiplied by c^2, that is,

$$Q = [m({}_{3}Li^{6}) + m\ ({}_{1}H^{2}) - m({}_{2}He^{4}) - m({}_{2}H^{4})]c^{2}.$$

Putting the given values in atomic mass unit (*u*), we have

$$Q = (6.015123\ u + 2.014102\ u - 4002603\ u - 4.002603\ u)c^{2}$$

$$= (0.024019\ u)c^{2}.$$

But $1\ u \times c^{2} = 931.5$ MeV.

$$\therefore \quad Q = 0.024019 \times 931.5 = 22.37 \text{ MeV.}$$

5. *Compute the Q-value of the reaction* $_{1}H^{3}$ ***(d, n)*** $_{2}He^{4}$. ***Given : mass of neutron = 1.008665 u, mass of deuterium***

(atom) =2.014102 u, mass of tritium (atom)=3.016049 a, mass of helium atom = 4.002603 a and 1 amu ≡ 931.5 MeV.

Solution: The reaction is

$$_1H^3 + {_1H^2} \rightarrow {_2He^4} + {_0n^1}.$$

Its Q value is given by

$$Q = [m(_1H^3) + m(_1H^2) - m(_2He^4) - m(_0n^1)]c^2$$

$$= (3.016049 \text{ u} + 2.014102\ u - 4.002603\ \text{u} - 1.008665\ \text{u})c^s$$

$$= (0.018883\ \text{u})\ c^2$$

$$= 0.018883 \times 931.5 = 17.6 \text{ MeV}.$$

6. *Compute the Q-value of the Rutherford's reaction :*

$$_7N^{14} + {_2He^4} \rightarrow {_8O^{17}} + {_1H^1}.$$

The atomic masses are : $_7N^{14}$ *= 14.003074 u,* $_2He^4$ *= 4.002603 u,* $_8O^{17}$ *= 16.99913l u and* $_1H^1$ *= 1.007825 u.*

Solution: The Q – value of the reaction is given by

$$Q = [m(_7N^{14}) + m\ (_2He^4) - m(_8O^{17}) - m(_1H^1)]c^2$$

$$= (14.003074 \text{ u} + 4.002603\ u - 16.999131\ \text{u} - 1.007825\ \text{u})\ c^2$$

$$= 18.005677\ u - 18.006956\ u)\ c^2$$

$$= (--0.001279\ u)\ c^2.$$

The energy equivalent of mass 1 u is 931.5 MeV, that is, lu × c^2 = 931.5 MeV.

$$\therefore \qquad Q = -\ 0.001279 \times 931.5 = -\ 1.19 \text{ MeV}.$$

Because the Q-value is negative, the reaction is "endothermic"

7. *Compute the Q-value of the Chadwick's reaction :*

$$_4Be^9 + {_2He^4} \rightarrow {_6C^{12}} + {_0n^1}.$$

The atomic masses are : $_4Be^9$ = 9.012182 u, $_2He^4$ = 4.002603 u, $_6C^{12}$ = 12.000000 u, neutron mass = 1. 008665 u and 1 amu is equivalent to 931.5 MeV of energy.

[Ans. 5.70 MeV.]

8. *Find the mass of $_6C^{14}$ (in amu) from the following nuclear reaction :*

$$_7N^{14} + {_0n^1} = {_6C^{14}} + {_1H^1} + 0.55 \text{ MeV}.$$

Given : mass of proton = 1.00758 amu, mass of neutron = 1.00898 amu, mass of $_7N^{14}$ = 14.00752 amu and 1 amu = 931 MeV.

Solution: The *Q*-value of the reaction is 0.55 MeV. We know that

$$Q = [m({_7N^{14}}) + m({_0n^1}) - m({_6C^{14}}) - m({_1H^1})]c^2.$$

Let the mass of $_6C^{14}$ be *mc* amu. Substituting the given values in the above equation, we have

$$(14.00752 \text{ amu} + 1.00898 \text{ amu} - mc - 1.00758 \text{ amu})\, c^2$$

$$= 0.55 \text{ MeV}$$

or$(14.00892 \text{ amu} - mc)c^2 = 0.55$ MeV.

Now 1 amu × c^2 = 931 MeV. Therefore, we can write

$$(14.00892 \text{ amu-}mc)c^2 = \frac{1 \text{ amu} \times c^2}{931} \times 0.55$$

or14.00892 amu-mc = 0.00059 amu

or $\qquad mc = 14.00892 - 0.00059 = 14.00833$ amu.

9. ***Calculate the minimum energy of a photon that will break the nuclide $_3Li^7$ into an α-particle and a tritron. The atomic masses are : $_8Li^7$ = 7.016004 u, $_2He^4$ = 4.002603 u, $_1H^8$ = 3.016050 u and 1 u × c^2 = 931.5 MeV.***

Solution: The equation for the given photo-disintegration process is

$$_3Li^7 + \gamma = \ _2He^4 + \ _1H^3$$
(photon) (α-particle) (tritron)

The *minimum* energy of the photon must be equal to the *rest* energy of the product particles minus the (rest) energy of $_8Li^7$. Let us write the above equation in terms of energy :

$$\gamma = [m(_2He^4) + m\ (_1He^3) - m(_3Li^7)]c^2$$
$$= (4.002603 \text{ u} + 3.016050 \text{ u} - 7.016004 \text{ u})\ c^2$$
$$= (0.002649 \text{ u})c^2$$
$$= 0.002649 \times 931.5 = 2.47 \text{ MeV}.$$

10. *Complete the following reactions and calculate the Q-value in each case. ($_4Be^9$ = 9012182 u. $_4Be^7$ = 7.016930 u, $_3Li^7$ = 7.016004 u, $_3Li^6$ = 6.015123 u, $_1H^1$ 1.007825 u, $_1He^4$ = 4.002603 u, $_0n^1$=1.008665 u, 1 u × c^2 = 931.5 MeV)*

(i) ***$_3Li^7$ (p, n) ?***

(ii) ***$_4Be^9$ (p, α) ?***

[Ans. *(i)* $_3Li^7$(p , n) $_4Be^7$, Q = – 1.645 MeV.

(ii) $_4Be^9$ (p, α) $_3Li^6$, *Q* = 2.125 MeV.]

4

Artificial Radioactivity

Artificial Nuclear Disintegration

Rutherford's Discovery of Artificial—Nuclear Disintegration—Discovery of Proton : The radioactive elements, in which natural disintegration of nucleus occurs, are the heavier elements. Rutherford was working for artificially disintegrating the nuclei of *lighter* elements by bombarding them with high-speed particles. He, in 1919, succeeded in disintegrating nitrogen nuclei by bombarding ordinary nitrogen gas with α-particles emitted from RaC′.

Rutherford's apparatus consisted of a long chamber with a side opening covered by a silver foil *F* (Fig.). A zinc sulphide (ZnS) screen was placed just outside the opening and a microscope *M* was placed for observing any scintillations occurring on the screen. The source of α-particles, *S*, was deposited on a metal plate placed inside the chamber. The distance of *S* from the ZnS screen could be varied. The chamber could be filled with different gases through the side-tubes T_1 and T_2.

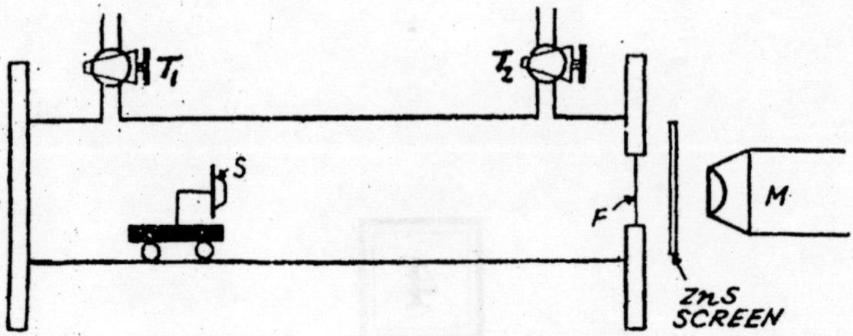

When the chamber was filled with oxygen or carbon-dioxide gas, no scintillations were seen on the screen. *But when it was ,felled with nitrogen, scintillations were observed on the screen,* even though this screen was shielded from the α-particle source S by the silver foil *F* which was thick enough to absorb all the α-particles. Hence Rutherford concluded that the scintillations were produced by some new particles which were more penetrating that the α-particles and were ejected from the nitrogen nuclei by collision with α-particles.

Since the scintillations were observable even when the distance of *S* from the screen was as long as 40 cm, the ejected particles had a range upto 40 cm. Magnetic deflection experiments indicated that these particles had a positive charge equal to the charge of an electron and a mass equal to that of a hydrogen nucleus (1.672×10^{27}kg). They were called as 'protons'.

Blackett, in 1925, took cloud chamber photographs of the above process. He obtained a straight thicker track of the incident α-particle,. and a long thinner track of the ejected proton together with a short, thick track of the recoiling nucleus. *There was no α-particle track after the collision, showing* that the α-particle had disappeared completely. From this it was concluded that the α-particle did not rebound after collision but was absorbed by the nitrogen nucleus, resulting into a new nucleus and the emission of a proton. The nuclear reaction for this process is

$$_7N^{14} + {}_2He^4 \rightarrow {}_8O^{17} + {}_1H^1$$

nitrogen α-particle oxygen proton
isotope

This reaction states that when a nitrogen nucleus ($_7N^{14}$) is hit by an α-particle ($_2He^4$), a proton ($_1H^1$) is ejected leaving a recoiling oxygen nucleus ($_8O^{17}$). Thus Rutherford transformed ordinary nitrogen into a rare isotope of oxygen. This was the first *artificial* nuclear transformation.

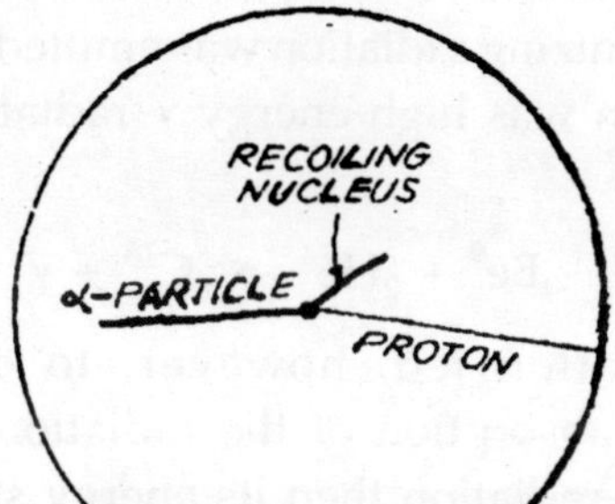

Following Rutherford historic experiment on artificial nuclear disintegration, Rutherford and Chadwick disintegrated other light elements by bombarding them with α-particles. They found that protons could be ejected out of the nuclei of all the light elements from boron to potassium (with the exception of carbon and oxygen).

In some cases the energy of the ejected protons was even greater than that of the bombarding α-particles. This result further verified that the protons were emitted due to nuclear disintegration, the extra energy being acquired due to the nuclear re-arrangement.

Later on, particles other than those emitted from a radioactive substance, such as neutrons, and artificially-accelerated charged particles (protons, deuterons) were used for producing nuclear disintegration.

Discovery of Neutron

In 1920, there were only three known material particles, namely electron, proton and α-particle, all charged. Rutherford had suggested the existence of a new particle having a mass roughly equal to the proton mass but carrying no charge. The particle was subsequently discovered and is now known as the neutron.

In 1930, Bothe and Becker found that when beryllium (or boron) was bombarded with α-particles, a highly penetrating but very poorly ionizing radiation was emitted. It was supposed that this radiation was high-energy γ–radiation produced by the reaction

$$_4Be^9 + {}_2He^4 \rightarrow {}_6C^{13} + \gamma$$

This supposition led, however, to difficulties. The measurement of absorption of the radiation in lead showed that if it was a γ–radiation then its energy should be about 7 MeV. This value was greater than the energy of any γ–radiation known at that time.

In 1932, Curie and Joliot investigated the radiation further by examining its effect on a paraffin block (a substance rich in hydrogen) placed between the beryllium and an ionisation chamber (Fig.). They found that the ionisation increased markedly by the presence of the block. They correctly concluded that protons (hydrogen nuclei) were being ejected from the paraffin by the radiation, and producing ionisation in the chamber. They assumed that the supposed γ–radiation on falling upon the paraffin, underwent Compton collisions with the hydrogen nuclei, which therefore recoiled and appeared as protons.

These protons were found to have energies of about 4.5 MeV. Calculations on this basis showed that each incident γ–ray photon must have had an energy of 55 MeV, a value about eight times higher than that deduced from the absorption measurements. Thus there was a serious difference between the values of the energy of the supposed γ–radiation given by the two methods.

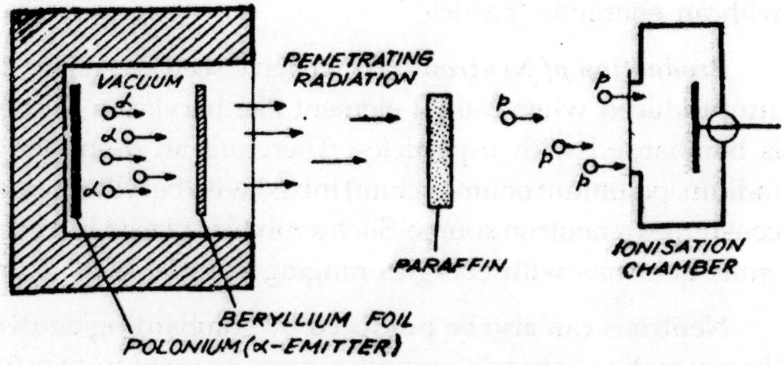

In 1932, Chadwick performed a series of experiments on the recoil of many other nuclei (for example nitrogen) when struck by this penetrating radiation. He found that if this radiation consisted of γ–ray photons, then the energy of the photons as obtained from experimental results varied with the nature of the recoiled nucleus. For example, protons ejected from paraffin had energies which required a γ–ray photon to have an energy of 55 MeV, while recoiling nitrogen nuclei had energies which required a γ–ray photon to have an energy of 90 MeV. This meant that the energy of the supposed γ–ray photon increased with the mass of the recoiling atom, which was contrary to the conservation of energy and momentum in Compton collisions.

Chadwick showed that these difficulties disappeared if the radiation coming from beryllium bombarded with α-particles is supposed to consist of 'particles' (instead of massless γ–ray photons) of mass nearly equal to that of a proton, but having no charge. He called these particles 'neutrons'. The nuclear reaction which produces these neutrons is

$$_4Be^9 + {}_2He^4 \rightarrow {}_6C^{12} + {}_0n^1$$

where $_0n^1$ is the symbol for the neutron. The penetrating nature of the neutrons follows from the absence of a charge, and the energies of the recoiling nuclei in Chadwick's experiments can be accounted for on the basis of collisions with an energetic 'particle.'

Production of Neutrons : As we have seen above, neutrons are produced when a light element like beryllium (or boron) is bombarded with α-particles. Therefore an α-emitter (like radium, polonium or americium) mixed with beryllium powder constitutes a neutron source. Such a mixture placed in a capsule emits neutrons with energies ranging upto 10 MeV or more.

Neutrons can also be produced by bombarding deuterium (heavy water) or beryllium with γ–rays obtained from artificial radioactive atoms like $_{11}Na^{24}$ or $_{51}SB^{124}$. Such sources are called 'photoneutron' sources and give practically monoenergetic neutrons.

Now a days the most powerful source of neutrons is the nuclear reactor in which the fission of heavy nuclei takes place.

Determination of Mass of Neutron : Since neutron is not a charged particle, its mass cannot be determined directly by deflecting it in electric or magnetic field. Chadwick determined the mass of neutron by measuring the maximum velocities of recoiling nuclei of hydrogen and nitrogen struck by neutrons. Suppose a neutron of mass m and velocity v suffers head-on collision with a stationary hydrogen nucleus of mass m_H. Let the velocity of the neutron after the collision be v', and the (maximum) velocity of the recoiling hydrogen nucleus be v_H. The equation of conservation of kinetic energy is

$$\frac{1}{2}mv^2 = \frac{1}{2}mv'^2 + \frac{1}{2}m_{H''H^2},$$

and the equation of conservation of momentum is

$$mv = mv' + m_H v_H.$$

Eliminating v' from these two equations, we get

$$v_H = \frac{2m}{m + m_H} v. \qquad \text{... } (i)$$

Again, if a neutron with the same velocity v collides with a stationary nitrogen nucleus of mass m_N, the (maximum) velocity imparted to the recoiling nitrogen nucleus is given by

$$v_N = \frac{2m}{m + m_N} v. \qquad \text{... } (ii)$$

Dividing eq. (*i*) by (*ii*), we get

$$\frac{v_H}{v_N} = \frac{m + m_N}{m + m_H}$$

m_N and m_H are known. If, therefore, v_H and v_N are found by measuring the maximum length of the cloud chamber tracks of the recoiling nuclei, the mass m of the neutron can be determined. Chadwick's result, although approximate because of errors in the determination of v_H and v_N, showed that the mass of the neutron is slightly larger than that of the proton.

Properties :

(i) The neutron is a fundamental constituent of the nuclei of all atoms (except hydrogen atom). It has a mass of 1.00898 amu or 1.675×10^{-27} kg which is slightly greater than that of a proton.

(ii) It is an uncharged particle. Therefore it cannot be accelerated to high velocities by means of electric fields as can be charged particles such as protons and electrons. For the same reason, the neutrons cannot be focussed by means of magnetic fields.

(iii) It is a highly penetrating particle and can pass through thick sheets of lead.

(iv) Being chargeless, the neutron produces practically no

ionisation in a gas, and hence no track in a cloud chamber.

(v) Being uncharged, a neutron can easily enter the nucleus of an atom. It can therefore produce a nuclear excitation or nuclear disintegration far more readily than almost any other particle. (Other particles carrying a charge have to overcome the strong electrostatic repulsion offered by the nucleus.)

(vi) If the probability of nuclear excitation or nuclear disintegration is small, the neutrons, on striking matter, are simply scattered by the atomic nuclei. On colliding with heavy nuclei, the neutrons are scattered with very little loss of energy. On colliding with light nuclei, however, the neutrons are slowed down in a few collisions. Light water (H_2O), heavy water (D_2O), paraffin wax, and carbon are very effective in slowing down neutrons. These substances are called 'moderators'. A slow neutron is more efficient in producing nuclear disintegration because it spends more time near a nucleus than a fast neutron and thus stands a greater chance of being captured.

(vii) Neutrons possessing energies of 1 MeV or more are known as *fast* neutrons. Those with energies below 1 eV are described as *slow* neutrons. The neutrons which have come into thermal equilibrium with a moderator at normal temperature and pressure are called *thermal* neutrons. Such neutrons have energies of approximately 0.03 eV.

Detection of Neutrons : The neutron, being a non-ionising particle, does not produce a track in a cloud chamber. Hence it cannot be detected by a G-M tube or by a cloud chamber. Indirect methods are, however, available for its detection.

Slow neutrons can be detected by means of a G-M tube or ionisation chamber filled with boron trifluoride gas (containing

$_5B^{10}$). The neutrons passing through the gas disintegrate boron nuclei which thus emit α-particles :

$$_5B^{10} + {_0n^1} \rightarrow {_3Li^7} + {_2He^4} \text{ (α-particle)}$$

The α-particles so produced cause ionisation of the gas and are thus detected.

Similarly, a G-M tube or ionisation chamber containing hydrogen will detect fast neutrons. When a fast neutron collides with a hydrogen nucleus, it imparts energy to the hydrogen nucleus which in turn produces ionisation in the surrounding gas and is detected.

Radioactive Decay of Neutron : A free neutron outside an atomic nucleus is *unstable* and decays into a proton, emitting an electron and an antineutrino. The reaction is

$$_0n^1 \rightarrow {_1H^1} \quad + \quad {_{-1}e^0} \quad + \quad \overline{v}$$

(neutron) (proton) (electron)
(antineutrino)

The half-life of the neutron has been estimated to be 12.8 min.

Uses of Neutrons :

(i) Neutrons are used in medicine, specially in the treatment of cancer.

(ii) Fast and slow neutrons are used for artificial disintegration of nuclei and producing radio-isotopes.

(iii) Slow neutrons are used in nuclear fission.

Positron : It is a positively-charged particle having the same mass and charge as an (negative) electron. Thus it is the *antiparticle* of the electron and is also called as 'positive electron'.

Positron was discovered by Anderson in 1932. Anderson was photographing the tracks of comic-ray electrons in a Wilson cloud chamber placed in a strong, magnetic field. He obtained a number of curved tracks showing that they were formed by charged particles of electronic mass and electronic charge. The direction of curvature of most of the tracks indicated a negative charge on the particles. Occasionally, however, a track was obtained whose direction of curvature was just opposite, thus indicating a positive charge on the particle producing it. The particle was named positron. It was the first antiparticle to be discovered.

The existence of positron was predicted by Dirac before its discovery by Anderson. It was later found that when hard γ–rays are absorbed by nuclei of atoms, the electrons and positrons are simultaneously produced (pair production). Positrons are also emitted by some artificially produced radioactive substances.

A positron has only an extremely short life, of the order of a micro-second. It is readily annihilated by combining with an electron from the surroundings.

Artificial Transmutation of Elements (or Artificial Disintegration of Nuclei): An artificial disintegration of nuclei is a process in which a nucleus is transformed into a different species by its reaction with an energetic particle or photon. The formation of the product nucleus is accompanied by the emission of one or more light particles, or photon. The whole process occurs very rapidly in a time of about 10^{-13} second or less. It is also known as artificial nuclear disintegration' or 'nuclear reaction'.

The nuclear reaction which was the first to be discovered by Rutherford is

$$ {}_7N^{14} + {}_2He^4 \rightarrow {}_8O^{17} + {}_1H^1 $$

(α-particle) (proton)

In this reaction a nitrogen nucleus hit by an α-particle is converted into an oxygen nucleus and emits a proton. For every nuclear reaction the total number of protons and neutrons is conserved as is the total mass *plus* energy.

Artificial Radioactivity

The experiments on the artificial disintegration of nuclei led to the discovery of artificial (or induced) radioactivity. In 1934, Curie and Jollot observed that when certain light elements, like boron and aluminum, were bombarded by α-particles, the resulting products of disintegration emitted positrons, and the emission persisted *even after bombardment by α-particles was stopped.*

The positron activity decayed exponentially with time just in the same way as natural radioactivity. This showed that as a result of α-particle bombardment, the stable elements were converted into unstable radioactive isotopes. *The phenomenon in which a stable element is converted into a radioactive isotope by an artificial disintegration is called 'artificial radioactivity'.* These artificially-produced radioactive isotopes have comparatively much shorter half-lives than the natural radioactive elements.

Thus when boron is bombarded by α-particles, it disintegrates emitting a neutron and forming an unstable radioactive isotope of nitrogen called radio-nitrogen :

$$_5B^{10} + {_2He^4} \rightarrow {_7N^{13}} + {_0n^1}.$$

The radio-nitrogen $_7N^{13}$ decays into $_6C^{13}$ by emitting a positron :

$$_7N^{13} \rightarrow {_6C^{13}} + {_{+1}e^0} \qquad \text{(positron)}$$

The half-life of $_7N^{13}$ is found to be about 14 min, and $_6C^{13}$ produced is a stable isotope of carbon.

In a similar manner, the bombardment of aluminium results in the production of unstable radio-phosphorus, which decays by positron-emission into a stable isotope of silicon :

$$_{13}Al^{27} + {}_{2}He^{4} \rightarrow {}_{15}P^{30} + {}_{0}n^{1}$$

$$_{15}P^{30} \rightarrow {}_{14}Si^{30} + {}_{+1}e^{0}$$

In general, artificial radioactive isotopes are produced by the bombardment of stable elements with accelerated charged particles (protons, deuterons, α-particles). By far the most useful particles for producing artificial radioactivity are the neutrons which are available in large numbers in a nuclear reactor. Hence artificially radioactive isotopes are now-a-days produced by placing elements inside a nuclear reactor. Practically all the elements can now be made artificially radioactive.

Some artificially radioactive isotopes emit positrons, while some others emit electrons. In some cases γ–ray emission also takes place together with positrons or electrons. Artificially radioactive isotopes which emit α-particles are much less common and occur mostly in transuranic elements (Z>92).

There is some correlation between the type of radioactivity of an artificially radioactive isotope and the means of its formation. The radioactive isotopes produced by (n, γ), (n, p), (n, α), and (d, p) reactions in which the neutron to proton ratio increases, emit electrons. As an example, we consider (n, α) reaction :

$$_{13}Al^{27} + {}_{0}n^{1} \rightarrow {}_{11}Na^{24} + {}_{2}He^{4}$$

$$_{11}Na^{24} \rightarrow {}_{12}Mg^{24} + {}_{-1}e^{0}$$

(electron)

Radioactive isotopes produced by (p, γ), (p , n), (α, n), (d, n) and (γ, *n*) reactions in which the neutron to proton ratio decreases, emit positrons. As an example, we consider (α, n) reaction :

$$_{7}N^{14} + {}_{2}He^{4} \rightarrow {}_{9}F^{17} + {}_{0}n^{1}$$

$$_{9}F^{17} \rightarrow {}_{8}O^{17} + {}_{+1}e^{0} \text{ (positron)}$$

Radio-isotope and their Applications : In addition to the naturally occurring radio-isotopes such as radium, hundreds of others have been made artificially. These isotopes have numerous applications in medicine, agriculture, industry and pure research. Many applications employ a special technique known as 'tracer technique'.

Tracer Technique : A small quantity of a radio-isotope is introduced into the substance to be studied and its path is traced by means of a G. M. counter. As an example, a leakage in an underground water-pipe can be detected by this method. A small quantity of radio-sodium Na^{24} (γ–ray emitter) is introduced into the pipe at its inlet. After the liquid has gone through the pipe, the ground around the leak will have larger γ–ray activity which can be detected by moving a G. M. counter on the ground. Similarly, in order to locate a blockage in an underground sewage pipe, a rubber ball having Na^{24} is introduced into the pipe. A G. M. counter above ground will give the position of the ball when it has come to rest.

Radio-isotopes can also be used in transporting *different* oils through underground pipe to distant places. When the type of the oil flowing through the pipe is changed, a small quantity of radio-isotope is mixed exactly at the position where the change takes place. Near the other end of the pipe line, a Geiger counter is placed which gives a signal when the radio-isotope passes.

In the field of medicine the tracer technique is employed in a number of ways. For example, the doctor can find out any obstruction in the circulation of the blood in the human body. He injects radio-phosphorous (P^{32}) into the blood of the patient and examines the movement of the blood by detecting radiations emitted by P^{32} by means of G. M. counter. He can thus locate clots of blood present in the body. In a similar way, the passage of a particular element in the body and the rate at which it accumulates in different organs can be studied. For example, phosphorus accumulates in bones, and iodine in thyroid gland. When the thyroid gland suffers with some disease, its rate of accumulation of iodine changes. To investigate it, radio-iodine (I^{131}) is given orally to the patient and the radiation emitted by his thyroid gland is measured externally by a G. M. counter at suitable intervals over the following 48 hours or so. An over-active or under-active (ailing) thyroid gland can thus be diagnosed.

In agriculture, the tracer technique is used to study the rate and direction of movement of an element in a plant. For this a radio-isotope of that element is injected in the ground near the plant. After a few days the plant is laid on a photographic paper to produce an auto-radiograph. The dark areas in the radiograph show the positions reached by the element. This technique gives valuable information regarding the optimum season for fertilising crops and for poisoning weeds.

In industry, the tracer technique is used for testing the uniformity of mixtures. For testing a chocolate mixture, a small quantity of a short-lived radio-isotope such as Na^{24} or Ma^{56} is added to the primary ingredients. Several different samples of the final product are then tested for radioactivity by means of a G. M. counter. If each sample gives the same counting rate, then the mixing has been uniform. This method can be used in mixing processes occurring in the manufacture of chocolate, soap, cement paints, fertilizers, cattle food and medical tablets.

The tracer technique is extremely sensitive in testing the sealing process in making envelopes for radio valves. A sample valve is filled with radio-krypton (Kr^{85}) and a G. M. counter is held outside the valve. The counter detects even an extremely poor leakage.

The tracer technique is also used in research to study the exchange of atoms between various molecules, and to investigate the solubility and vaporisation of materials.

Besides the uses employing the tracer technique there are hundreds of other uses of radio-isotopes in various fields.

Medical Uses : The radiations given out by some radio-isotopes are very effective in curing certain diseases. For example, radio-cobalt (Co^{80}) is used in the treatment of brain tumour, radio-phosphorus (P^{32}) in bone-diseases and radio-iodine (I^{131}) in thyroid cancer. The radiations, besides destroying the ailing tissue, also damage the healthy tissue and hence a careful control over the quantity administered is necessary.

Bacteria and other disease-carrying organisms can be destroyed by irradiating them with γ–rays. The process is used to sterilise medical instruments, plastic hypodermic needles, packets of antibiotics, and hospital blankets; whereas heat sterilisation would damage them. A portable source of γ–rays for sterilisation is radio-cobalt (Co^{60}).

X-ray photography in medical diagnosis can be replaced by γ–ray photography with advantage. The γ–ray source (radio-isotope) is compact and needs no power supply.

Agricultural Uses : Radiations from certain radio-isotopes are used for killing insects which damage the food grains. Certain seeds and canned food can be stored for longer periods by gently exposing them to radiations.

Better yields of milk from cows, and more eggs from hens have been obtained on the basis of information gained by mixing radio-isotopes with their diet.

Radio-isotopes are also employed for determining the function of fertilizer in different plants. Thus the agricultural yield is increased.

Certain seeds, when exposed to feeble radiation, develop into different varieties of plants. For example, new and exciting colours have been given to some of the flowering plants.

Industrial Uses : There are many different uses to which radio-isotopes are put in industry. By γ–ray photography we can find out wearing of cutting tools and lathes, and can locate internal cracks in stones. We can check any non-uniformity in the thickness of a sheet by β– or γ–absorption measurements. The sheet is made to run continuously between a radio-isotope (emitting β– or γ–rays) and a counter. A change in the counting rate indicates a variation in the thickness of the sheet. The output from the counter may be used to correct the machinery which is rolling the sheet as soon as a variation is detected, and thus the thickness is automatically kept constant. This method is used as a thickness control in the manufacture of paper, plastic, metal sheet, etc.

The same method can be used to check sealed cigarette packets whether they are full or if one or more cigarette is missing. The packets are placed on a conveyer belt running between a radio-isotope and a counters. An empty or partially-filled packet gives a higher counting rate due to less absorption of radiation than with a completely filled packet. The increase in counting rate can be converted into an electronic signal which knocks the incomplete packet off the belt.

A radioactive isotope together with a fluorescent material (such as ZnS) becomes a weak source of light. Such sources are used for providing light in coal mines, and for painting watches.

Carbon Dating : The radio-carbon (${}_6C^{14}$) is continuously being produced in the atmosphere by neutron bombardment of nitrogen :

$${}_7N^{14} + {}_0n^1 \rightarrow {}_6C^{14} + {}_1H^1$$

The radio-isotope ${}_6C^{14}$ has a half-life of 5600 years and decays back to nitrogen by electron emission :

$${}_6C^{14} \rightarrow {}_7N^{14} + {}_{-1}e^0.$$

In the course of time an equilibrium has been reached in which the rates of formation and decay of $_6C^{14}$ are equal. Hence the ratio of $_6C^{14}$ to stable carbon $_6C^{12}$ in all atmospheric carbon dioxide is constant.

All *living* matter (plant or animal) by one process or another exchanges its carbon with atmospheric carbon (in the form of carbon dioxide). As long as the exchange is taking place, the ratio of $_6C^{14}$ to $_6C^{12}$ in the plant or animal is the same as that in the atmosphere. After the death of a plant or animal the process of exchange stops, so that the concentration of $_6C^{14}$ in it decreases due to radioactive decay. The ratio of $_6C^{14}$ to $_6C^{12}$ remaining at a certain time after death, can be found by measuring the activity of a carbon sample taken from the dead body. From this ratio the time since death can be calculated. By this method, ages ranging from about 600 years to 20,000 years can be estimated. The method is known as 'carbon-14 dating'.

Carbon-14 dating is applicable for dating dead vegetation (such as wood, paper, papyrus, etc.) and animal matter (such as hair, wool, etc.). For instance, it has been used for wooden coffins containing Egyptian mummies, which were found to be 3500 years old.

Transuranic Elements : The heaviest naturally occurring element is uranium (*U*) which has an atomic number Z = 92. Since 1940, elements having an atomic number greater than 92 have been *artificially* produced in the laboratory by the bombardment of certain heavy nuclei with appropriate particles. These are called 'transuranic elements' (Z>92). The transuranic elements ranging from Z=93 to Z=103 are the following :

1. *Neptunium, Np (Z = 93)* : When uranium $_{92}U^{238}$ is bombarded with *slow* neutrons, a new isotope $_{92}U^{239}$ is produced with the emission of γ-rays (radioactive capture) :

$$_{92}U^{238} + _0n^1 \rightarrow _{92}U^{239} + \gamma$$

The product nucleus $_{92}U^{239}$ is radioactive and decays by electron emission into $_{93}Np^{239}$, an isotope of the first transuranic element neptunium :

$$_{92}U^{239} \rightarrow _{93}Np^{239} + _{-1}e^0 + \overline{v}$$

2. *Plutonium, Pu (Z = 94)* : The Neptunium isotope $_{93}Np^{239}$ is itself radioactive and has a half-life of 23 days. It decays by electron emission into $_{94}Pu^{239}$, an isotope of the second transuranic element plutonium :

$$_{98}Np^{239} \rightarrow _{94}Pu^{239} + _{-1}e^0 + \overline{v}$$

$_{91}Pu^{239}$ is very stable. It decays by α-emission into $_{92}U^{235}$ with a half-life of 24,400 years :

$$_{94}Pu^{239} \rightarrow _{92}U^{235} + _2H_e^4$$

The production of $_{94}Pu^{239}$ is carried on a large scale for generating nuclear power.

Other transuranium elements can be made by bombarding the appropriate heavy nucleus with accelerated α-particles, deuterons, or with bare nuclei of lighter atoms. They are named as below :

(3) Americium, Am (Z = 95)

(4) Curium, Cm (Z = 96)

(5) Berkelium, Bk (Z = 97)

(6) Californium, Cf (Z=98)

(7) Einsteinium, E (Z = 99)

(8) Ferminm, Fm (Z=100)

(9) Mendelevich, Mv (Z = 101)

(10) Nobelium, No (Z=102)

(11) Lawrencium, Lw (Z = 103)

Each one of the transuranium elements is found to have several isotopes. Elements beyond *E* (Z=99) have so short half-lives that they cannot be isolated in weighable quantities.

In general, the half-lives of transuranium elements are very short compared with the age of the earth ($\approx 4 \times 10^9$ years). Even if the elements existed among the natural ores when the earth was formed, they would have disappeared long ago. This is why they do not occur in nature. On the other hand, the half-life of uranium is about the same as the age of the earth so that a considerable quantity still remains.

PROBLEM

1. *10 mg of carbon from living material produces 200 counts per minute due to a small fraction of the radio-carbon* ${}_6C^{14}$. *A piece of ancient wood of mass 10 mg is found to give 50 counts per minute. Find the age of the wood assuming that the* ${}_6C^{14}$ *content of the atmosphere has remained unchanged. Half-life of* ${}_6C^{14}$ *is 5700 years.*

[**Ans.** 11400 years]

5

Particle Detectors

Nuclear Radiation Detectors

Particle Detectors : Nuclear radiation (or particles) such as $\alpha-$ $\beta-$, and $\gamma-$rays, cannot be observed directly. They are detected by the secondary effects which they produce in the materials through which they pass. There are three such effects, and a number of detecting instruments are based upon them. These effects are ionisation, photographic action and fluorescence.

Ionisation : A charged particle like $\alpha-$particle, $\beta-$particle, or proton, passing through matter produces ions by collisions with the molecules of the material. This ionisation is the basis of many detecting instruments such as electroscope, spark counter, ionisation chamber, Geiger-Muller counter, cloud chamber and bubble chamber. The various instruments differ in the material within which the ionisation is produced, and in the way in which the ionisation is observed or measured.

Electroscope : It is the simplest instrument used for the detection of ionisation produced in air by nuclear radiation. It consists of two gold leaves, one fixed in position, the other free to rotate. When the leaves are charged to a positive potential (say) with respect to earth, they are separated out due to mutual electrostatic repulsion. If the air surrounding the leaves is now exposed to a radioactive source *R*, it becomes ionised so that positive ions and electrons are produced.

The positive ions move towards the earthed case, and the electrons towards the leaves. The electrons neutralise some of the positive charge on the leaves which therefore come closer. The decrease in the separation of leaves is a measure of the intensity of the ionising radiation, specially of the α–rays which alone produce strong ionisation.

Spark Counter : It is another simple device for detecting α–particles. It consists of a tungsten wire *A* and a flat tungsten plate *B*, with an air gap of about 1 mm in between. A potential difference of about 6 kV, which is only very slightly less than that required to produce sparking, is applied across the gap. When an α–source is placed near the gap, sparks begin to pass across the gap. This is because the α–particle passing through the gap ionises the air strongly and a spark is initiated. The counting of spark visually gives roughly the number of α–particles passing through the gap. Alternatively, the current that passes in a spark produces a voltage-pulse in a resistance *R*, which is fed into an electronic scalar which records the exact number of α–particles.

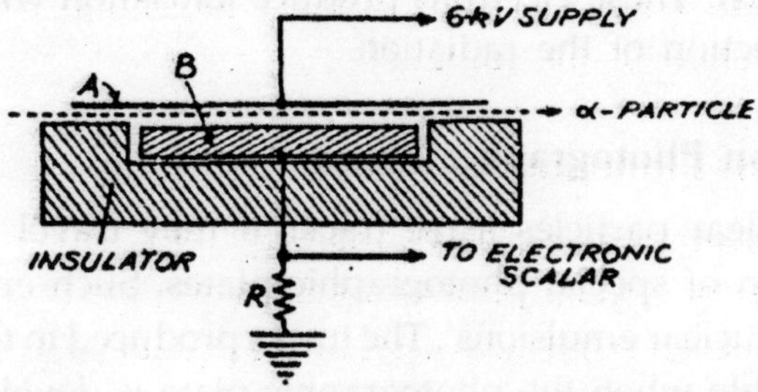

The spark counter can give the range of α–particles in air by finding the maximum distance between source and gap for spark to be produced, It responds well to α–particles, but it is practically insensitive to β–particles and photons (γ–rays, X-rays, etc.).

Ionisation Chamber and G-M Counter : These are the most popular instruments for detecting nuclear radiations. In them the ionisation is produced in a gas. The positive ions and the electrons so-produced are separated and collected by means of an electrostatic field. This results in electric pulses which are amplified, measured and counted electronically. ʻ

Cloud Chamber : In a cloud chamber the ionisation is produced in a gas *supersaturated* with some vapour, and the track of ions is made visible by the condensation of vapour upon them.

Babble Chamber : In a bubble chamber the ionisation is produced in a *superheated liquid* and the track of tons is made visible by the formation of bubbles upon them.

In both, the cloud chamber and the bubble chamber, the tracks are illuminated and photographed automatically.

Many of the above instruments' are used to detect not only the charged particles (such as α– and β–particles), but also the uncharged radiation such as γ–rays and X-rays which do not produce enough ionisation directly. These radiations are made to fall on suitable materials from which they eject electrons (by photoelectric, pair production, or Compton scattering processes). These electrons produce ionisation which leads to the detection of the radiation.

Effect on Photographic Plates

Nuclear particles leave tracks if they travel through the emulsion of special photographic plates. Such emulsions are called 'nuclear emulsions'. The tracks produced in them become observable when the photographic plate is developed. Many important studies of cosmic rays and nuclear particles have been made with the aid of nuclear emulsions.

Fluorescence : Nuclear radiation, especially α–particles, produce flashes of light when they fall upon a screen coated with a fluorescent material like zinc sulphide or barium platino-cyanide. These flashes of light are called 'scintillations'. The observation and counting of these scintillations provide us a means of detecting nuclear particles.

Rutherford was the first to use fluorescence for detecting α–particles. He made α–particles to struck a ZnS screen and observed visually the scintillation produced. Crookes in 1903 designed a simple instrument called 'spinthariscope' on this basis. It consists of a small radioactive source *R* and a fluorescent screen *S*, together with an eye-piece *E* for viewing the scintillations. The number of scintillations counted per unit time gives directly the number of α–particles falling upon the screen. The screen is comparatively insensitive to β– and γ–rays. It has the disadvantages that scintillations can only be seen in a dark room, and that visual counting is limited to about 60 scintillations/min.

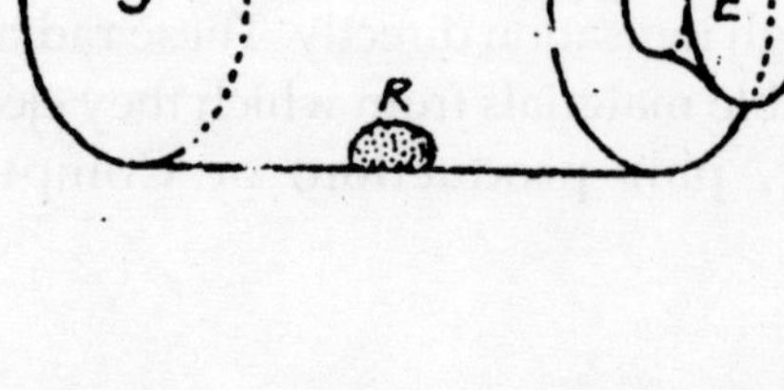

Since 1947, various liquids and solids have been found in which scintillations can be produced by β– and γ–rays also, as well as by neutrons (through an indirect process). In modern instruments, the scintillations are converted into amplified current pulses by a photo-multiplier tube, and are counted electronically. Such an instrument is called a 'scintillation' counter.

Gas-filled Detectors : When nuclear radiation passes through a gas enclosed between two electrodes, it ionises the gas molecules. If a potential difference is applied across the electrodes, the positive ions move toward the negative electrode and the electrons toward the positive electrode. Thus pulses of electric current are produced which detect end measure the ionising radiation. Such detectors are called 'gas-filled detectors' and include *ionisation chamber, proportional counter* and *Geiger-Mueller counter*. Each one of them has its own condition of ionisation under which it can work. To understand this, let us consider the variation of ionisation current with the potential difference applied across the electrodes. The curve I corresponds to the ionisation current produced by a charged particle of lower energy; while the curve II corresponds to a particle of higher energy.

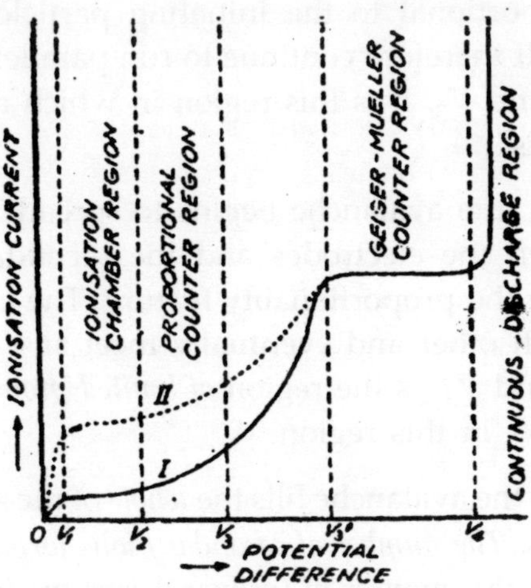

If there is no p.d. across the electrodes, the ions will recombine and there will be no current. As the p.d. is increased to a few volts, some of the positive ions and electrons reach the opposite electrodes so that a small ionisation current is obtained. At a certain p.d. V_1, *all* the ions formed in the gas reach the electrodes and a saturation current results. As the p.d. is further increased, the current stays constant until a p.d. V_2 is reached. Thus between V_1 and V_2 *the ionisation current is independent of the applied p.d.*, and is proportional to the number of (primary) ions formed in the gas, and hence proportional to the energy of the ionising particle. If the ionising particle is of higher energy, a larger saturation current is obtained, as shown by the dotted curve. In the region between V_1 and V_2 the ionisation chamber works.

When the p.d. is increased above V_2 the (primary) ions formed by the charged particles gain so much energy that they themselves begin to ionise the gas molecules by collision. The secondary ions so formed, in turn, initiate further ionisation, and so on. In this way an avalanche of ionisation is produced and the current increases. Upto a p. d. V_3, the number of secondary ions formed remains proportional to the number of (primary) ions produced initially. Hence the output current pulse is proportional to the initiating particle energy. The curves I and II therefore continue to run parallel in the region between V_2 and V_3. It is this region in which a proportional counter works.

Above V_3, the avalanche begins to spread rapidly in the space between the electrodes and the secondary ionisation begins to lose the proportionality feature. The curves I and II approach each other and eventually meet at V_3'. The region between V_3 and V_3' is the region *of limited proportionality*. No detector works in this region.

Above V_3' the avalanche fills the *whole* of the space between the electrodes. *The number of secondary ions formed now become independent of the, number of (primary) ions produced initial,* so that the curves I and II become identical. The region above V_3' is the region of Geiger-Mueller counter. It ends at a p. d V_4, above which continuous discharge starts.

Ionisation Chamber : An ionisation chamber is used for detecting mainly the α– and β–rays, and to compare the activities of sources emitting these rays. Basically,, it consists of a chamber filled with a gas like air or argon at normal pressure. It is fitted with a pair of electrodes E_1 and E_2 between which a constant p. d. is maintained. It also carries a window W made of nylon or mica, 0.002 mm thick, and coated with graphite to make it conducting.

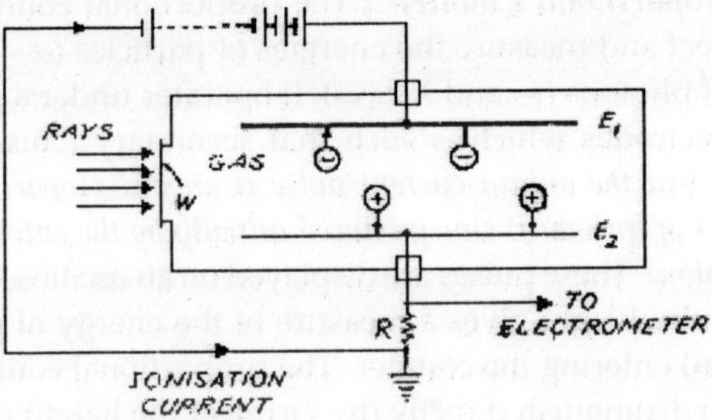

When the rays to be detected enter the chamber through the window, they ionise the gas between the electrodes. The positive ions and the electrons so produced move in opposite directions towards the electrodes so that an ionisation current ($\approx 10^{-12}$ amp) flows in the external circuit. This current is measured by measuring the p. d. established by it in a high resistor R by means of an electrometer.

The p. d. across the electrodes is so selected that *all* the ions collect at the electrodes and there is *no* secondary ionisation. Under this condition the (saturation) ionisation current is proportional to the number of pairs of ions and hence to the number of particles entering the chamber. Therefore, the activities of the sources can be compared by comparing their ionisation currents.

An ionisation chamber is much less sensitive to β–particles (in comparison to α–particles) because β–particles produce fewer pairs of ions in their passage through the chamber. It is too insensitive for γ–rays.

For detecting γ–rays, an ionisation chamber of thick wall made of high atomic-number material (Pt, Bi) is employed. The γ–rays impinging on the *walls* of the chamber eject high-speed electrons which produce ionisation in the gas.

The ionisation chamber does not give a count of individual particles, but an average effect of a large number of particles.

Proportional Counter : The proportional counter is used to detect and measure the energies of particles (α– and β) and also of photons (γ– and X-rays). It operates under a p. d. across the electrodes which is such that secondary ionisation takes place, but *the output current pulse is always proportional to the number of (primary) ions produced initially by the entering particle (or photon).* These pulses are displayed on an oscilloscope screen. The pulse height gives a measure of the energy of particle (or photon) entering the counter. The proportional counter is thus able to distinguish directly (by virtue of the height of the pulse produced) between α– and β–particles and γ–rays since they differ so much in ionising powers.

Geiger-Mueller Tube (G-M Tube) : A Geiger-Mueller tube is the most versatile and useful of the instruments used for detecting and measuring the energies of α, β–, γ–, and X-rays. It is sufficiently sensitive to detect individual α– and β–particles. It is widely used in industry and medicine to locate radio isotopes, and in the laboratory for comparing the activity of radioactive specimen.

One of the simplest form of a G. M. tube is shown in Fig. It consists of a metallic cylindrical cathode tube enclosed in a glass envelope. A fine tungsten wire anode is stretched along the axis of the tube. The rays to be detected enter the tube through the glass envelope.

A slightly different form of G-M tube is used for α–particles and the less energetic β–particles which are unable to penetrate the glass envelope. It is known as 'end-window' G. M. tube and is shown in Fig. In it the cylindrical cathode tube carries a thin mica window to let the particles enter. The wire anode, instead of extending through the length of the tube, terminates in a point.

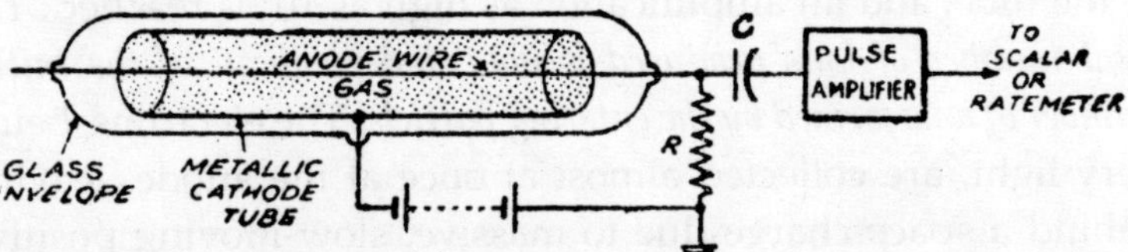

(a)

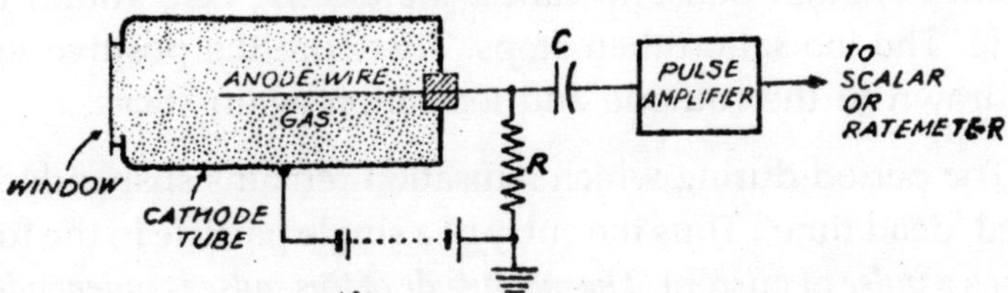

(b)

The tube contains a monatomic gas such as argon or neon at a pressure of about 10 cm of mercury, together with a small trace of a halogen such as chlorine or bromine. (The halogen acts as a 'quenching agent'). A potential difference of several hundred volts is applied between the anode wire and the cathode tube.

When an ionising particle enters the G-M tube, it ionises some of the argon atoms, forming positive ions and free electrons. Because of the shape of the electrodes, the electrostatic field is radial and is very strong near the anode wire. Any free electrons in this region are rapidly accelerated to cause further ionisation. This process is cumulative, and a large 'avalanche' of electrons is produced. Besides this, some of the free electrons colliding the argon atoms merely *excite* them. The excited atoms return to their normal state emitting light photons. If a photon is absorbed by another excited atom, the atom is ionised releasing more electrons which produce further avalanches. Since photons are unaffected by the electric field, they can go in any direction.

Hence the avalanche spreads rapidly in the *entire volume* of the tube, and an amplification as high as 10^8 is reached. *The total number of ions produced is now 'independent' of the initial number of ions formed by the entering particle.* The electrons, being very light, are collected almost at once at the anode, leaving behind a space-charge due to massive, slow-moving positive ions. In a short time ($\approx$ 1 micro-second) the space-charge becomes enough dense to cancel the electric field round the anode. The ionisation then stops. Therefore the positive ions are drawn to the cathode and ionisation again starts.

The period during which ionisation remains suspended is called 'dead time'. Thus the entry of a single particle in the tube causes a *pulse* of current. *The magnitude of this pulse is independent of the energy and nature of the ionising particle.* The current pulse causes a corresponding voltage pulse across the resistance R in series with it. This voltage pulse is fed to a pulse amplifier through a capacitor C. The amplified pulse is finally passed on either to a scalar or to a ratemeter. The scalar records the arrival of each individual pulse separately, and so gives the exact number of particles entering the G-M tube in a given time-interval. The ratemeter records the number of pulses in a given time, *i.e. it* gives the average rate at which the particles enter the tube. The combination of G-M tube and the recorder is known as 'G-M counter'.

Because the magnitude of the pulse produced does not depend upon the nature or energy of the entering particle, a G-M tube does not distinguish between the arrival of an α–particle and a β–particle. (A proportional counter does it).

However, the particles can be distinguished by finding what absorber placed in their path will prevent them from entering the G-M tube.

Self-quenching Action of the G-M Tube : The G-M tube must produce a single pulse due to the entry of a single particle. It should not then give any spurious pulses, but should recover quickly to record the next entering particle. Unfortunately, the positive argon ions which eventually strike the cathode gain electrons from it and become neutral argon atoms which are left in an excited state. These excited atoms return to the normal state with the emission of photons, and these photons cause avalanches and hence spurious pulses.

These unwanted pulses are avoided by the presence of a poly-atomic or diatomic gas, such as bromine, in the tube. The positive argon ions moving *slowly* toward the cathode collide many times with the bromine molecules and transfer their charge to them. Thus only neutral argon atoms reach the cathode. On the other hand, now bromine molecular *ions* reach the cathode where they gain electrons to become neutral bromine molecules in an excited state. However, the excited bromine molecules lose their excitation energy by dissociation into bromine atoms, *and not by photon emission,* and thus no spurious pulses are produced. In due course the bromine atoms recombine to form bromine molecules. Thus bromine *quenches* the discharge, and the tube is ready to receive the next particle within about $\frac{1}{10000}$ th of a second.

Operating Voltage for G-M Tube : It is important to operate a G-M tube at the correct p.d. A plot of counting rate against the p.d. applied is shown in Fig. For low p.d.'s, only the most energetic particles produce detectable pulses. As the p.d. is increased, the count rate increases. In this region the magnitude of the pulse depends upon the energy of the entering particle ; and only those particles are recorded which give detectable pulses. Beyond a certain *threshold* p.d. there is a range of p.d.'s, *LM*, in which the count rate is almost constant. This is known as the plateau region'. In this region all pulses are of the same magnitude, and every entering particle is being recorded. This is the right range of p.d.'s in which the G-M tube must be operated. For a halogen-quenched tube this range is about 400-500 volts. Beyond the plateau region, the quenching action becomes ineffective, The tube may then go into a 'continuous discharge" and be damaged.

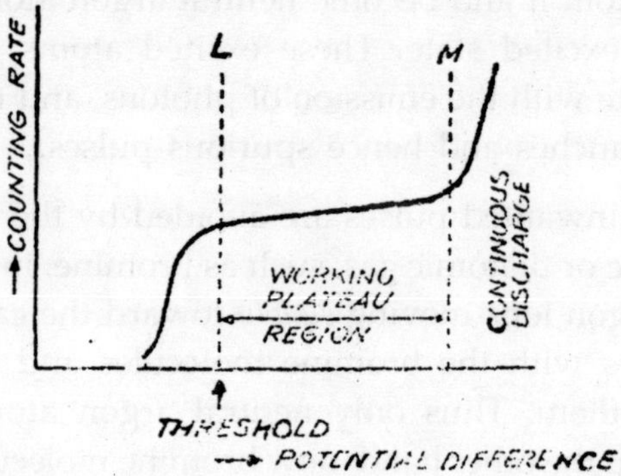

Wilson's Cloud Chamber : C.T.R. Wilson, in 1911, invented a cloud chamber to detect the presence of radioactive particles, to identify them and to measure their energy.

Principle : If we suddenly increase the volume of air saturated with water vapour, the temperature of air falls sharply and it becomes super-saturated with water vapour. The vapour is condensed in the form of small droplets on the dust particles present in the air and a fog appears. If there be no dust particles in the air, the vapour will not be condensed and the fog will not appear.

If, however, we ionise the 'dust-free' air by some ionising radiation (like α– or β–particles), then *the vapour is condensed on the ions produced in the air*. Thus, by seeing the condensation of water vapour in dust-free air, the presence of β–and β–particles can be detected.

Construction : Wilson's cloud chamber is shown in Fig. It consists of a cylindrical metallic chamber *C* fitted with a glass top *G* and carrying a movable piston *P*. The space between the piston *P* and the glass cover *G* is filled with *dust-free* air saturated with water-vapour. Intense light is sent into the chamber through a window. Just above the chamber is fitted a camera. Near the chamber is placed some radioactive substance at the end of a rod.

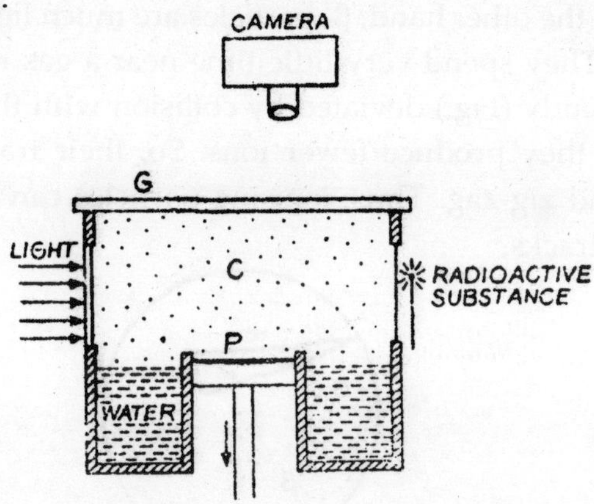

Working : The α or β–particles emitted by the radioactive substance enter the chamber. Each particle goes on producing ions by ionising the air molecules on its path. To mark the path of the particle the piston *P* is dropped down which leads to a sudden expansion of the saturated air in the chamber. Hence the air is cooled and becomes super-saturated with water vapour. Now the vapour condenses in the form of drops on the ions formed along the path of the particle. Thus the path of the particle is marked as a trail of droplets which is the track of the particle. It is photographed by the camera. The process is automatic. As soon as a particle enters the chamber, the piston drops down, the shutter of the camera opens and the track of the particle is photographed. Immediately after this, the particle track is removed by applying an electric field across the air space. Now the chamber gets ready for the next photograph.

Uses—Detection of Ionising Particles : Different ionising particles (α, β, etc.) produce different tracks in Wilson cloud chamber. For example, α–particles are heavier and slow-moving. Therefore, they remain comparatively longer in the vicinity of the molecules on their way, ionise a larger number of them and soon lose their energy. Hence the α–particle tracks in the cloud chamber are thick, short and almost continuous (Fig.). On the other hand, β–particles are much lighter and fast moving. They spend very little time near a gas molecule and are frequently (Fig.) deviated by collision with the molecules. As such, they produce fewer ions. So, their tracks are thin, dotted and zig-zag. Thus, ionising particles can be identified by their tracks.

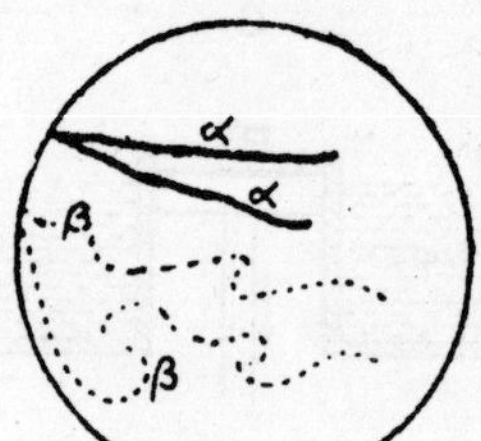

Determination of Momentum, Energy and the Sign of the Charge of Ionising Particles : For this, the cloud chamber is placed in a known magnetic field vertically down the chamber so that the tracks of the charged particles become curved (Fig.). The measurement of the curvature of the path determines the momentum and energy of the particle while the direction of curvature determines the sign of the charge.

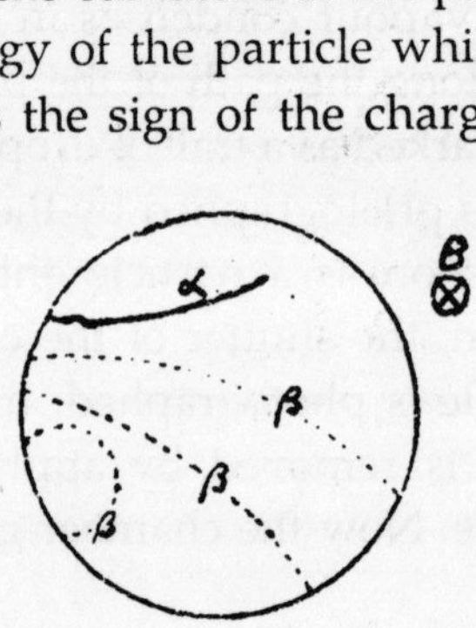

We know that a particle of mass m, charge q and velocity v moving perpendicularly to a magnetic field of flux density B describes a circular path of radius r, given by

$$\frac{mv^2}{r} = qvB.$$

Therefore, the momentum p of the particle is

$$p = mv = rqB. \qquad \text{... } (i)$$

Hence, by measuring the radius r of the circle, the momentum p of the particle can be determined.

If the particle velocity v is comparable to the velocity of light c, then the kinetic energy K of the particle is:

$$K \simeq pc.$$

So, knowing the momentum p, the kinetic energy of the particle can be calculated.

Discovery of New Particles

The discovery of particles positron and meson and the study of cosmic rays could be possible by the Wilson's cloud chamber.

The cloud chamber is not suitable for γ-rays and X-rays because they produce very little ionisation in the gas.

Automatic Cloud Chamber : Blackett and Occhialini placed a cloud chamber C between two G–M tubes G_1 and G_2, connected in coincidence (Fig). An ionising particle which passes through both these tubes necessarily passes through the cloud chamber C. The pulse produced by the coincidence circuit operates a thyratron which, in turn, triggers off the sudden expansion of the gas in C and also an spark illumination for photography of the track. After a trad, has been photographed, the pulse also starts an electric motor which resets the apparatus and makes it ready to record fresh particle.

Bubble Chamber : The bubble chamber is a device, invented by Glaser in 1952, to photograph the tracks of charged particles, it is almost the exact (Fig.) inverse of cloud chamber. When a liquid is subjected to a pressure higher than normal, it can be held in the liquid state at a temperature *above* its normal boiling point. The liquid then becomes *superheated* at the moment when the pressure, is suddenly reduced, and has a strong tendency for boiling. If the liquid is very pure and free from dust, this superheated state can be maintained for a few seconds, after which it starts boiling. If an ionising particle passes through the superheated liquid, it leaves a trail of ions behind it. These ions cause immediate boiling and, acting as centres, form a track of bubbles. This track of bubbles can be readily illuminated and photographed.

A simple form of bubble chamber is shown in Fig. It consists of a heavy-walled pyrex bulb filled with a low boiling-point liquid like liquid-propane or liquid-hydrogen. The liquid is compressed by admitting air through a pressure-regulating device, and is maintained at a proper temperature by a thermostat- controlled oil-bath placed round the pyrex bulb. The air is then allowed to escape very rapidly so that the liquid becomes highly superheated at normal pressure and remains so for a few seconds. During this period, a charged particle is passed through the liquid which leaves a track of bubbles behind it. Light is flashed and the bubbles are photographed simultaneously from different angles.

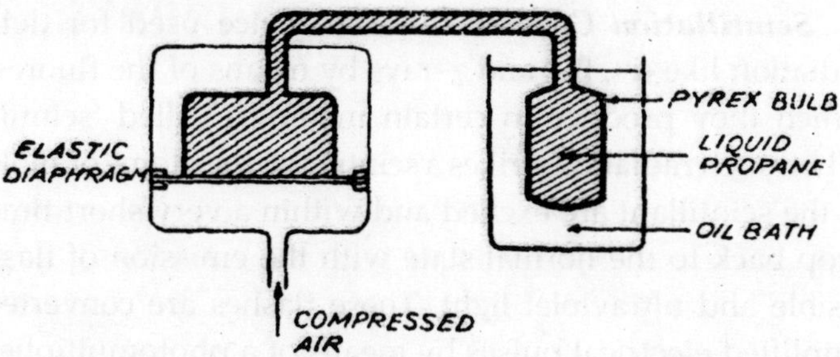

The bubble chamber is more sensitive, more efficient and more compact than a cloud chamber and gives sharped tracks.

Photographic Emulsions

When an ionising particle (such as proton, α–particle or electron) travels through a photographic plate emulsion, it leaves a track in the emulsion which appears when the plate is developed. This provides us a technique for detecting ionising particles.

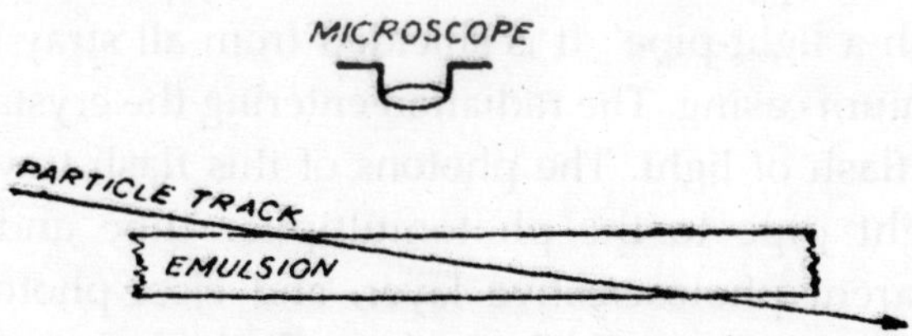

Special photographic emulsions are prepared for this purpose. They are called 'nuclear emulsions'. They are thicker, and have a large size and density of silver bromide grains than an ordinary optical emulsion. The tracks which are left by the ionising particles passing through these emulsions are short (a few mm) so that they are studied under a microscope (Fig.). Counting of the individual tracks gives a measure of the number of particles entering the emulsion, and the study of the structure of the track gives information about the mass, charge and energy of the particle. The method has the advantage of giving a permanent record of the events studied.

Scintillation Counter : It is a device used for detecting radiation like α–, β–, and γ–rays by means of the fluorescence which they produce in certain materials called 'scintillants'. When such radiation strikes a scintillant, the atoms or molecules of the scintillant are excited and within a very short time they drop back to the normal state with the emission of flashes of visible and ultraviolet light. These flashes are converted into amplified electrical pulses by means of a photomultiplier tube and are then recorded electronically.

The selection of the scintillant depends upon the radiation to be detected. For α-particles, zinc sulphide (ZnS) in the form of a thin crystal is used. For β–particles, a crystal of anthracene or naphthalene is commonly employed, while for γ–rays a crystal of sodium iodide (NaI) containing a trace of thallium is used.

One type of scintillation counter is shown in Fig. A suitable scintillant crystal is fitted to the end of a photo multiplier tube through a light-pipe'. It is shielded from all stray light by an aluminium casing. The radiation entering the crystal produces a tiny flash of light. The photons of this flash travel through the light pipe to the photomultiplier tube and fall on a transparent photosensitive layer, and eject photo electrons. The tube has several electrodes called 'dynodes' to which progressively higher potentials are applied.

The photoelectrons are pulled to the dynode 1 where a number of secondary electrons are emitted for each primary photo electron. These, in turn, are pulled to dynode 2 where the electrons are further multiplied by secondary emission. The process is repeated at all the dynodes and finally a highly amplified electrical pulse emerges at the anode. These pulses are fed to an electronic system where they are counted.

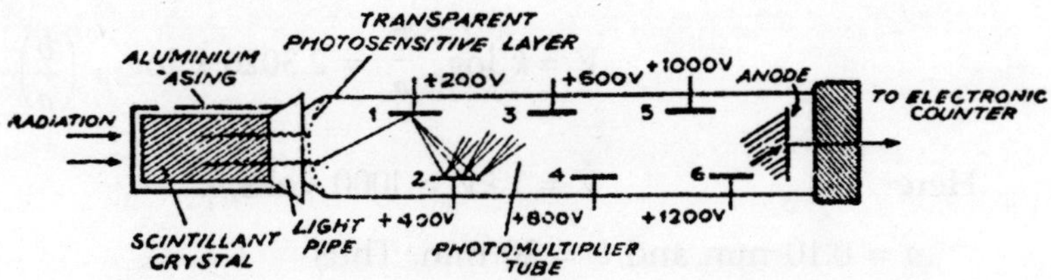

The scintillation counter has two main advantages over other counters such as the G-M tube counter. Firstly, the time of flight of the electrons through the photo multiplier tube is so small that as many as 10^6 particles per second can be counted. This is about 200 times the rate possible with G-M counters. Secondly, the efficiency of the scintillation counter in counting γ-rays is much higher.

Besides this, the magnitude of the output pulse from the photo multiplier is proportional to the energy of the particle or γ-photon entering the scintilla crystal. Hence a scintillation counter can be used as an energy spectrometer.

PROBLEMS

1. ***A halogen-quenched G-M tube operates at 1 kV and has a wire of diameter 0.20 mm. The radius of the cathode is 20 mm and the tube has a guaranteed life-time of 10^9 counts. What is the maximum radial field and how long will the counter last if it is used on the average for 30 hours per week at 3000 counts per minute ?***

Solution: Let a be the radius of the wire (anode) and b that of the cylindrical counter tube (cathode). The *radial* field will be maximum at the surface of the wire, that is, at a distance a from the center. It is given by

$$E = \frac{k}{a},$$

where k is a constant. The expression for the p.d. across the tube is given by

$$V = k \log_e \frac{b}{a} = 2.3026\, k \log_{10} \left(\frac{b}{a}\right).$$

Here $V = 1\ kV = 1000$ volts,

$a = 0.10$ mm and $b = 20$ mm. Thus

$$1000 = 2.3026\, k \log_{10} 200$$

$$\therefore \quad k = \frac{1000}{2.3026 \times 2.3010} = 188.7 \text{ volts.}$$

The (maximum) field at the wire surface is

$$E = \frac{k}{a} = \frac{188.7 \text{ volts}}{0.10 \text{ mm}} = 1887 \text{ volts/mm}$$

Suppose the life-time of the tube is N weeks. Then

$$N \times 30 \times 60 \times 3000 = 10^9$$

so that $$N = \frac{10^4}{54} = 185 \text{ weeks.}$$

If 1 year is taken to be of 50 weeks, then

$$N = \frac{185}{50} = 3.7 \text{ years.}$$

6

The Radiations

Natural Radioactivity

The atomic nuclei of certain elements and their salts, such as Uranium, Thorium, Radium, Polonium, etc., emit spontaneously invisible radiation which penetrates through opaque substances, ionises gases and affects photographic plates. Such elements are said to be "radioactive", and the spontaneous emission of radiation is known as "natural radioactivity".

Rutherford studied the effect of electric and magnetic fields on the radiation emitted by different radioactive substances (Fig.). He observed that the radiation has three types of rays : one which deflect toward the negative plate, second which deflect toward the positive plate, and the third which remain undeflected in the electric field. These are called 'alpha' rays' (α–rays), 'beta rays' (β–rays) and 'gamma rays' (γ-rays) respectively. α– and β–rays are actually streams of particles ; hence they are called α– and β–particles. Thus, α-particles are positively charged, α-particles are negatively charged and γ-rays are electrically neutral.

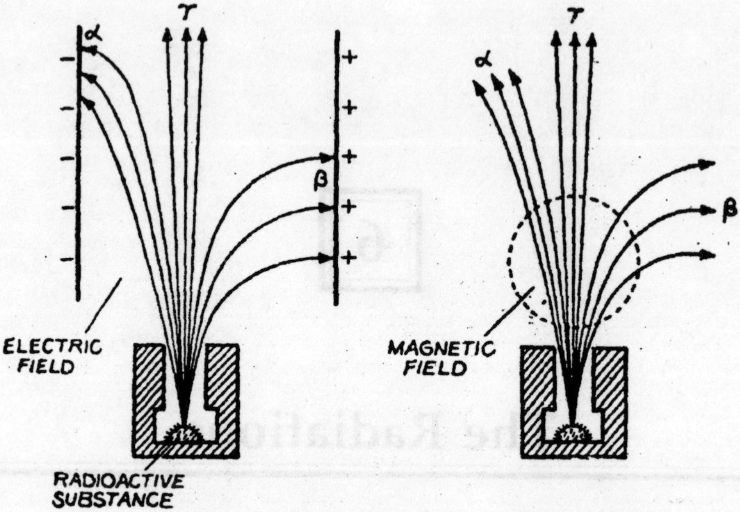

The same conclusion was drawn by passing the radiation through a magnetic field. The magnetic field in Fig. is perpendicular to the plane of paper, directed inward. In this field, α-particles are deflected to the left and β–particles to the right and move on circular paths, whereas γ-rays continue moving on their initial path. According to Fleming's left-hand rule, α-particles are positively charged and β–particles are negatively charged.

No radioactive substance emits both α- and β–particles simultaneously. Some substances emit α-particles, and some other emit β–particles. γ-rays are emitted along with α- and β–particles.

Properties of α-particles : An α-particle has a positive charge of 3.2×10^{-19} coulomb and a mass of 6.645×10^{-27} kg which are the same as the charge and mass of a helium nucleus. In fact, an α-particle is a helium nucleus and is represented by $_2He^4$. The main properties of α-particles are the following:

$$1000 = 2.3026\, k \log_{10} 200$$

1. α-particles are deflected in electric and magnetic fields, and the direction of deflection indicates that they are positively charged. Their small deflection (compared to β–particles) indicates that they are comparatively heavier particles.

2. The velocity of α-particles is much less than the velocity of light (less than 1/10th part), and α-particles emitted by different elements have somewhat different velocities. For example, the velocity of α-particles emitted by Uranium-1 is 1.4×10^7 meter/ second and that of particles emitted by Thorium-C′ is 2.2×10^7 meter/second.

3. The range of α-particles in air (the distance travelled by an α-particle in air at NTP) varies from 2.7 cm for particles from Uranium-1 to 8.6 cm for particles from Thorium C′. In general, the range varies with the radioactive substance, and the nature and pressure of the medium.

4. α-particles can penetrate through matter but their penetrating power is small. They are stopped by only 0.1 mm thick aluminium sheet. Their penetrating power is only 1/100th of that of β–particles and 1/10,000th of that of γ-rays.

5. They cause intense ionisation in a gas through which they pass. Their ionising power is 100 times greater than that of β–rays and 10,000 times greater than that of γ-rays. Their tracks in a cloud chamber are continuous.

6. They are scattered when passing through thin foils of gold or mica. While most of the particles scatter through small angles, some of them scatter through very large angles, even greater than 90°.

7. They produce fluorescence in substances like zinc sulphide and barium platinocyanide. When a particle strikes a fluorescent screen, a scintillation (flash of light) is observed. We can count the number of α-particles by counting the number of scintillations.

8. Because of their high emitting velocity, α-particles are used for bombarding the nuclei in the transmutation of one element into other.

9. They affect a photographic plate, though feebly.

10. They produce heating when stopped.

11. They cause incurable burns on human body.

Properties of β-particles : A β–particle has a negative charge of 1.6×10^{-19} coulomb which is the charge of electron. Actually, β–particles are fast-moving electrons. They have the following main properties :

1. β–particles are deflected by electric and magnetic fields and the direction of deflection shows that they are negatively charged. Their deflection is much larger than the deflection of α-particles. This shows that β-particles are much lighter than the α-particles.

2. Their velocity is from 1% to 99% of the velocity of light (only in velocity, β-particles differ from cathode rays). There is enough variation in the velocities of β-particles emitted by the same radioactive substance. This is why enough dispersion is found in these particles in electric and magnetic fields (Fig.).

3. Since the velocity of β-particles is of the order of the velocity of light, their mass increases with their velocity.

4. The β-particles emitted by the same radioactive substance has a continuous distribution of kinetic energy between zero and a certain maximum value,

and this maximum value is different for different substances. Hence the range of β-particles is not definite (while the range of α-particles is definite).

5. The penetrating power of β-particles is about 100 times larger than the penetrating power of α-particles. They can pass through 1 mm thick sheet of aluminium.

6. β-particles produce ionisation in gases. Their ionising power is much smaller than that of α-rays, because the mass of β-particle is much less. As β-particles cannot produce ionisation continuously, their tracks in cloud chamber do not appear to be continuous.

7. They are readily scattered while passing through matter.

8. They produce fluorescence in calcium tungstate, barium platinocyanide, etc.

9. They affect photographic plate more than the α-particles.

Properties of γ-rays : Like X-rays, the γ-rays are electromagnetic waves (or photons) of very short wavelength, ≈ 0.01 Å, which is about 1/100th part of the wavelength of X-rays. The important properties of γ-rays are as follows :

(1) γ-rays are not deflected by electric and magnetic fields. This indicates that they have no charge.

(2) They travel with the speed of light.

(3) They are most penetrating. They can pass through 30 cm thick iron sheet.

(4) They are diffracted by crystals in the same way as X-rays.

(5) They ionise gases, but their ionisation power is very small compared to that of α- and β-particles.

(6) They produce fluorescence.

(7) They affect photographic plate more than β-particles.

(8) Though there is much similarity between X-rays and γ-rays, yet their sources of origin are different. X-rays are produced by the transition of electrons in an atom from one energy level to another energy level, *i.e.*, it is an atomic property ; whereas γ-rays are produced from the nucleus, *i.e.*, it is a nuclear property.

(9) These rays are absorbed by substances and give rise to the phenomenon of pair-production. When a γ-ray photon strikes the nucleus of some atom, its energy is converted into an electron and a positron (positively-charged electron), and its own existence is extinguished:

$$\underset{(\gamma\text{-photon})}{h\nu} \rightarrow \underset{(\text{electron})}{e^-} + \underset{(\text{positron})}{e^+}$$

α-particles are doubly-charged Helium Ions : α-particles are emitted from radioactive atomic nuclei. Their deflections in magnetic and electric fields indicate that they are positively-charged. Experiments have shown that an α-particle carries a charge + 2e, which is numerically twice the charge of electron. Further, from the determination of the ratio of charge to mass, the mass of an α-particle has been calculated to be equal to the mass of a helium nucleus. From this we may conclude that α-particles may be doubly-charged helium ions, *i.e.*, helium nuclei. Rutherford and Royds, in 1909, showed experimentally that α-particles are in fact helium nuclei.

The experimental arrangement is shown in Fig. A small quantity of radon gas (radioactive substance) was sealed in a thin-walled glass tube *A*. This tube was placed in a thick-walled glass tube *B*. A capillary tube *C* provided with two electrodes was sealed on *B*. Tubes *B* and *C* were highly evacuated and the arrangement was allowed to stand for a few days.

The α-particles emitted by the radon passed through the thin walls of the tube *A* and collected in tube *B*. The gas collected in tube *B* could be forced into the capillary tube *C* by introducing mercury in *B*. After about a week enough gas was accumulated and forced into *C*. A discharge was then passed through the gas and the emitted light was examined by a spectroscope which clearly showed spectral lines of helium. The α-particles collected in the tube *B* had caught outer electrons and had become helium atoms. This spectroscopic evidence proves conclusively that α-particles are helium nuclei.

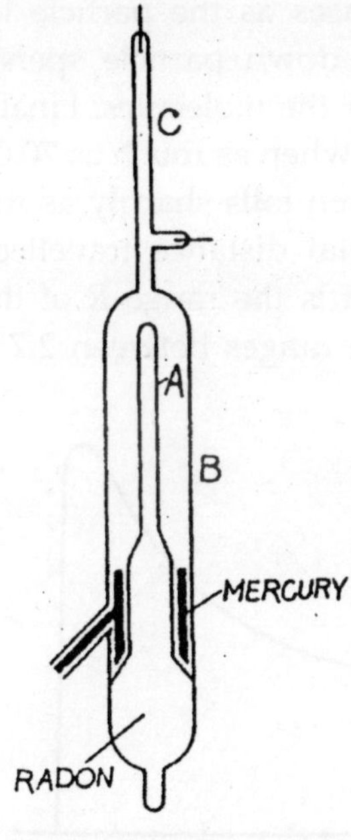

Detection of α-particles : α-particles strongly ionise air (or any other gas), and cause fluorescence in certain materials. Any of these properties may be utilised for the detection and counting of α-particles.

Two important α-particle detectors are spark counter (based on ionisation) and scintillation counter.

Range of α-particles :The distance through which an α-particle travels in a substance before coming to rest is called the 'range' of the particle in that substance. The α-particles from any one isotope are all emitted with approximately the same energy and have a well-defined range which is characteristic of that isotope.

When an α-particle passes through matter say air, it gradually loses its energy chiefly in inising the air molecules. At the start of its path, the particle may produce 20,000-30,000 ion pairs per cm of air. (This number can be found by measuring the density of the particle track obtained in a cloud chamber). The ionisation increases as the particle loses speed. This is because the slowing-down particle spends more and more time in the vicinity of the molecules. Finally, the unionisation reaches a maximum, when as much as 70,000 ion pairs per cm are produced, and then falls sharply as the particle comes to a stop (Fig.). The total distance travelled by the α-particle before coming to rest is the range *R* of the particle. Most α-particle sources have ranges between 2.7 and 8.0 cm in air.

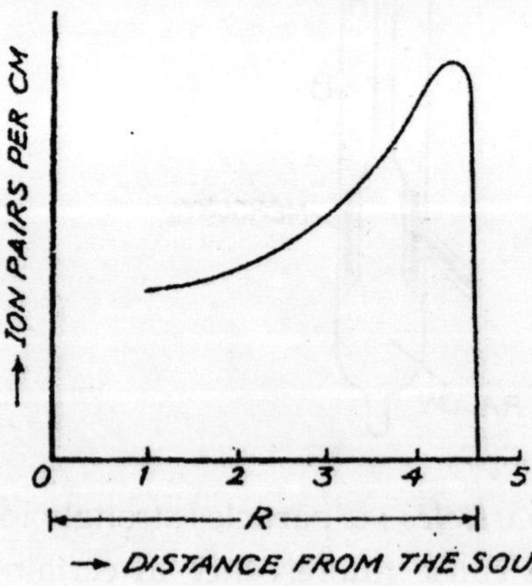

Range-Energy Relationship

The range of an α-particle in a substance is related to its initial kinetic energy. The energy lost by the particle per unit path in the substance is called the 'stopping power' $S(E)$ of the substance :

$$S(E) = -\frac{dE}{dx}$$

The stopping power varies with the energy of the particle, and the range of the particle is given by

$$R = \int_{E_0}^{0} \frac{dE}{S(E)} = \int_{E_0}^{0} \frac{dE}{-dE/dx},$$

where E_0 is the initial kinetic energy.

The stopping power $S(E)$ of a substance can be determined by measuring (e. g, by magnetic deflection) the energy of the particles after they have travelled a certain distance in the substance. When the energy loss is deter, mined for different initial velocities, the range in the substance can be deduced from the above relation as a function of the initial energy. A graph between range in air and initial energy is shown in Fig.

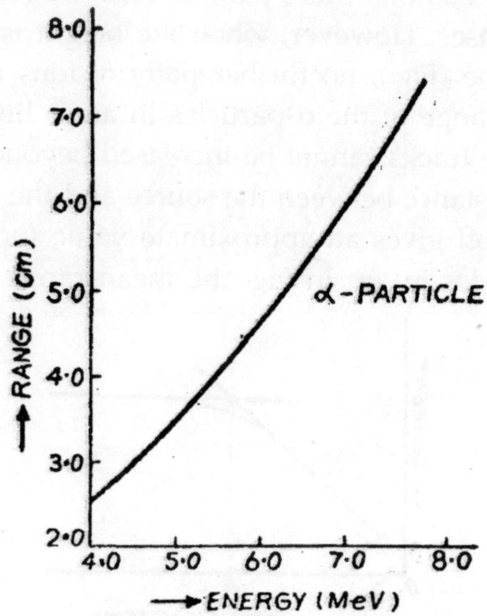

Experimental Determination of Range of α-particles: Experimentally, the range of α-particles in air can be measured using either a spark gap or a vertical ionisation chamber of adjustable height (Fig.). A radioactive source of α-particles is placed on a small platform inside the chamber whose top can be adjusted. A suitable constant potential difference is applied across the chamber, and the resulting ionisation current is measured by means of an electrometer.

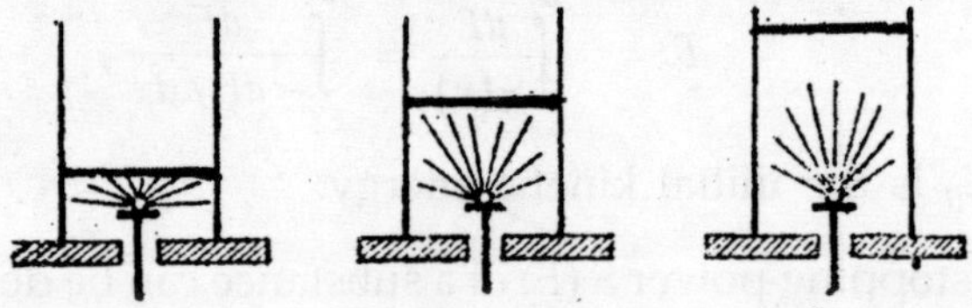

As the top of the chamber is raised, the current at first increases with the height of the chamber and then becomes constant (Fig.). In the beginning when the height is very small (Fig.), the tracks of the α-particles inside the chamber are short so that only few pairs of ions are produced.

As the height of the chamber is increased (Fig.) the tracks are lengthened so that more pairs of ions are formed and the current increases. However, when the height is greater than a certain value (Fig.), no further pairs of ions are produced because the range of the α-particles in air is limited and the lengths of the tracks cannot be increased beyond this limited range. The distance between the source and the top when the curve levels off gives an approximate value for the range of these α-particles in air. In Fig. the mean range is *OR*.

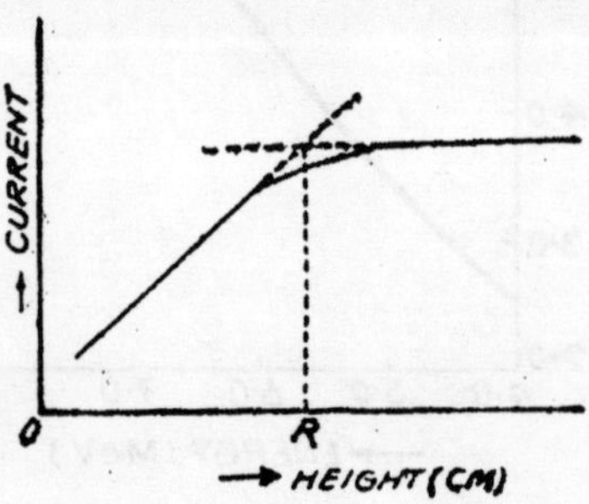

All the α-particles from a given radioactive source have the same range and hence the same energy. A close study, however, shows that α-particles often fall into a few close but sharp energy groups. This is known as the 'fine structure' of the α-rays. It gives the clue that there are different energy levels within the nucleus, *i.e.* the nucleus may exist in one or more excited states above the ground state. The existence of these excited states also explain the origin of λ rays from a radioactive nucleus.

Geiger-Nuttall Law

Different α-emitters emit α-particles of different energies, and hence of different ranges. It has been found that the α-emitters giving the higher-energy particles have the shortest half-lives (or largest decay constants), and vice-versa.

Geiger and Nuttall obtained an empirical relation between the decay constant λ of an α-emitter and the range *R* (in air) of the α-particles emitted by it. This relation which applies to members of a particular radioactive series is

$$\log \lambda = a + b \log R,$$

where *a* and *b* are constants for the given radioactive series. This law can be derived by quantum mechanics.

A few short-lived α-emitters, thorium *C*′ and radium *C*′, emit α-particles of unusually long range (10-12 cm) occasionally. Such a long-range α-particle is emitted from the parent nucleus when it is in an excited state, so that it carries not only its normal energy but also the excitation energy of the parent nucleus.

Radioactive Disintegration

Radioactive substances emit spontaneously either α-particles or β-particles, and some times γ-rays also. It is due to the disintegration of the nuclei of the atoms of the radioactive substance. This spontaneous emission is called 'radioactive disintegration' or 'radioactive decay'.

Rutherford-Soddy Theory of Radioactive Disintegration: Rutherford and Soddy studied the radioactive decay and formulated a theory which is based on the following laws :

(i) The radioactive emission is characteristic of the isotope, it varies from one isotope to another of the same element.

(ii) The emission occurs spontaneously and cannot be speeded up or slowed down by physical means such as change of pressure or temperature.

(iii) The disintegration occurs at random and which atom will disintegrate first is simply a matter of chance.

(iv) The rate of disintegration of a particular substance (i.e. number of atoms disintegrating per second) at any instant is proportional to the number of atoms present at that instant.

Let N be the number of atoms present in a radioactive substance at any instant t. Let dN be the number that disintegrates in a short interval dt. Then the rate of disintegration is $—dN/dt$, and is proportional to N *i.e.*,

$$-\frac{dN}{dt} = \lambda N.$$

where λ is a constant for the given substance and is called 'decay constant' (or 'disintegration constant' or 'radioactive constant' or 'transformation constant').

The above relation can be written as

$$\frac{dN}{F} = -\lambda \ dt.$$

Integrating, we get

$$\log_e N = -\lambda t + C,$$

where C is the integration constant. To determine C, we apply the initial conditions. Suppose there were N_0 atoms in the beginning *i.e.* $N = N_0$ at $t = 0$. Then

$$\log_e N_0 = C$$

$$\therefore \quad \log_e N = -\lambda t + \log_e N_0$$

$$\text{or} \quad \log_0 \frac{N}{n_0} = -\lambda t$$

$$\text{or} \quad \frac{N}{N_0} = e^{-\lambda t}$$

$$\text{or} \quad \boxed{N = N_0\, e^{-\lambda t}.}$$

This equation shows that the number of atoms of a given radioactive substance decreases exponentially with time (*i.e.* more rapidly at first and slowly afterwards). This is the Rutherford-Soddy law of radioactive decay.

The graph between the number of atoms left in a substance and the time is shown in Fig.

Statistical Nature of Radioactive Decay : Measurements of the rate of decay of a radioactive substance are purely statistical averages based on measurements made with a large number of atoms. Every atom in a sample of radioactive substance has a certain probability of decaying, but we cannot know which atom will actually decay in a particular time-interval.

Furthermore, an atom has no memory of the past. Suppose a particular atom has a probability of decay of 1 in 10^6 in a given length of time.

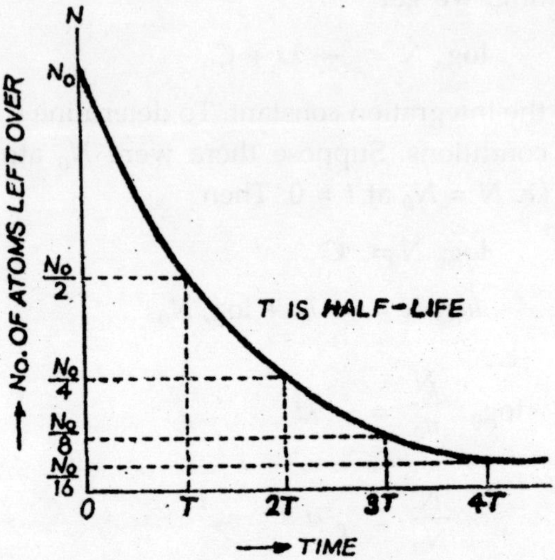

After a thousand years, if it has not decayed, it still has the same probability of decay. Thus for an atom the decay probability per unit time is constant until it actually decays.

Half-life : The atoms of a radioactive substance undergo continuous decay so that their number goes on decreasing. The time-interval T in which the mass of a radioactive substance, or the number of its atoms, is reduced to half its initial value is called the 'half-life' of that substance. The half-life of a radioactive substance is constant, but it is different for different substances. It can be read from the graph in Fig.

Relation between Half-life and Decay Constant : Let N_0 be the number of atoms present in a radioactive substance at time $t = 0$, and N the number at a later time t. Then, by Rutherford-Soddy law, we have

$$N = N_0 e^{-\lambda t},$$

where λ is the decay constant for the substance.

Now, at $t = T$ (half-life period), $N = \frac{1}{2} N_0$. Therefore,

$$\frac{N_0}{2} = N_0\, e^{-\lambda T}$$

or $$\frac{1}{2} = e^{-\lambda T}$$

or $$e^{\lambda T} = 2$$

or $$\lambda T = \log_e 2$$

or $$\boxed{T = \frac{\log_e 2}{\lambda} = \frac{0.693}{\lambda}.}$$

This is the relation between half-life and decay constant.

The half-life of a radioactive substance cannot be changed by any physical or chemical means. The half-life of an isotope of Lead ($_{82}Pb^{214}$) is 26.8 minutes. If this isotope forms some compound by chemical combination, even then its half-life will be 26.8 minutes.

The half-life of natural radioactive substances and their isotopes vary from a fraction of a second to hundreds of millions of years. The half-life of an isotope of Polonium ($_{84}Po^{214}$) is 10^{-5} second, whereas the half-life of Uranium ($_{92}U^{238}$) is 4.5×10^9 years.

The determination of half-lives is very useful for geologists for estimating the ages of mineral deposits, rocks and earth. We can also calculate how long our present stock of radioactive substances will last.

Average Life (or Mean Life) of Radioactive Atom : The atoms of a radioactive substance are continuously disintegrating. Which atom will disintegrate first is a matter of chance. Atoms which disintegrate in the beginning have a very short life and those which disintegrate at the end have the longest life. The average life of a radioactive atom is equal to the sum of the life-times of all the atoms divided by the total number of atoms.

Relation between Mean-life Time and Decay Constant : Suppose N_0 is the total number of atoms at $t = 0$, and N the number remaining at the instant t. Suppose a number dN of them disintegrates between t and $t + dt$. As the interval dt is small, we may assume that each of these dN atoms had a life-time of t seconds. Thus the total life-time of dN atoms is equal to $t\ dN$.

Since the disintegration process is a statistical one, any single atom may have a life from 0 to ∞. Hence the sum of the life-times of all the N_0 atoms is

$$\int_{t=0}^{t=\infty} t\, dN.$$

Dividing it by N_0, the total number of atoms originally present, we get the average life-time $\overline{T}$ of an atom. Therefore

$t = \infty$

$$\overline{T} = \frac{1}{N_0} \int_{t=0}^{t=\infty} t\, dN.$$

Now, from the disintegration law, we have

$$N = N_0 e^{-\lambda t}$$

or

$$dN = -N_0 \lambda e^{-\lambda t} dt$$

Substituting in the above expression, we get

$$\overline{T} = \frac{1}{N} \int_{t=0}^{\infty} t\left(N_0\ {}^{-\lambda t}\ dt\right) = \lambda \int_0^{\infty} t\, e^{-\lambda t}\ dt.$$

Integrating by parts, we get

$$\overline{T} = \lambda \left[\frac{t\, e^{-\lambda t}}{-\lambda}\right]_0^{\infty} - \lambda \int_0^{\infty} \frac{e^{-\lambda t}}{-\lambda}\, dt$$

$$= \left[-te^{-\lambda t}\right]_0^{\infty} + \int_0^{\infty} e^{-\lambda t}\, dt$$

The first term, on substituting the limits becomes zero. Therefore

$$\overline{T} = \int_0^\infty e^{-\lambda t}\,dt = \left[\frac{e^{-\lambda t}}{-\lambda}\right]_0^\infty$$

$$= \left[0 - \frac{1}{-\lambda}\right] \qquad [\because e^{-\infty} = 0]$$

$$= \frac{1}{\lambda}.$$

Thus, the mean-life $\overline{T}$ of a radioactive atom is equal to the reciprocal of its disintegration constant.

Relation between Mean-life and Half-life : The mean-life time $\overline{T}$ of a radioactive atom is different from the half-life *T*. We have seen that

$$\overline{T} = \frac{1}{\lambda}$$

and
$$T = \frac{0.693}{\lambda}.$$

$$\therefore \qquad \overline{T} = \frac{T}{0.693} = 1.44\,T.$$

Thus mean life is longer than half-life. The reason is that the last few atoms of a radioactive substance may last for a very long period of time.

Radioactive Series

Practically all natural radioactive elements lie in the range of atomic numbers from Z=83 to *Z*=*92*. The nuclei of these elements are unstable and disintegrate by ejecting either an α-particle or a β-particle. The ejection of an α-particle lowers the mass number *A* by 4, and atomic number Z by 2. The ejection of a β-particle has no effect on mass number but increases the atomic number by 1. The atomic number is characteristic of an element and a change in it implies the formation of an atom of a new element. The new atom so formed is also radioactive and further disintegrates into another new atom, and so on. Thus a series of new radioactive elements is produced by successive disintegrations until a stable element is obtained. Such a series is called a 'radioactive series'.

There are four important radioactive series : (*i*) Uranium series, (*ii*) Thorium series, (*iii*) Actinium series and (*iv*) Neptunium series.

Uranium Series : In this series the parent element is Uranium with atomic number Z = 92 and mass number A = 238 ($_{92}U^{238}$). The sequence starts by the loss of one α-particle giving the product $_{90}Th^{234}$, an isotope of thorium. This is followed by the emission of two β-particles which brings its nuclear charge to the original value 92, producing an isotope of Uranium, $_{92}U^{234}$. This, in turn, emits an α-particle thus producing another isotope of Thorium $_{90}Th^{230}$ which then emits another α-particle and becomes Radium $_{88}Ra^{226}$.

The end-product of this series, after the emission of five more α-particles and four more β-particles, is radium-lead ($_{82}Pb^{206}$), which is indistinguishable chemically from ordinary lead. It is a stable isotope of lead.

Several isotopes occurring in the series, such as $_{84}Po^{218}$, $_{83}Bi^{214}$, and $_{83}Bi^{210}$, have been found to decay in an alternative way also, as shown by dotted arrows in Fig.

Thorium Series : The parent element of this series is Thorium with Z = 90 and *A* = 232 ($_{90}Th^{232}$). It goes through a series of transformations in many respects similar to the Uranium series, and end with an stable isotope of lead ($_{82}Pb^{208}$).

Actinium Series : The parent element in this series is an isotope of Uranium called Actino-Uranium ($_{92}U^{235}$). The end product is again an stable isotope of lead ($_{82}Pb^{207}$).

Neptunium Series : With the discovery of the unstable transuranium element, another radioactive series was traced out.

This is called the 'Neptunium series' after its longest-lived member Neptunium. Its origin can be traced back to Plutonium. It does not end in a stable isotope of lead, but in the stable isotope of Bismuth ($_{83}Bi^{209}$).

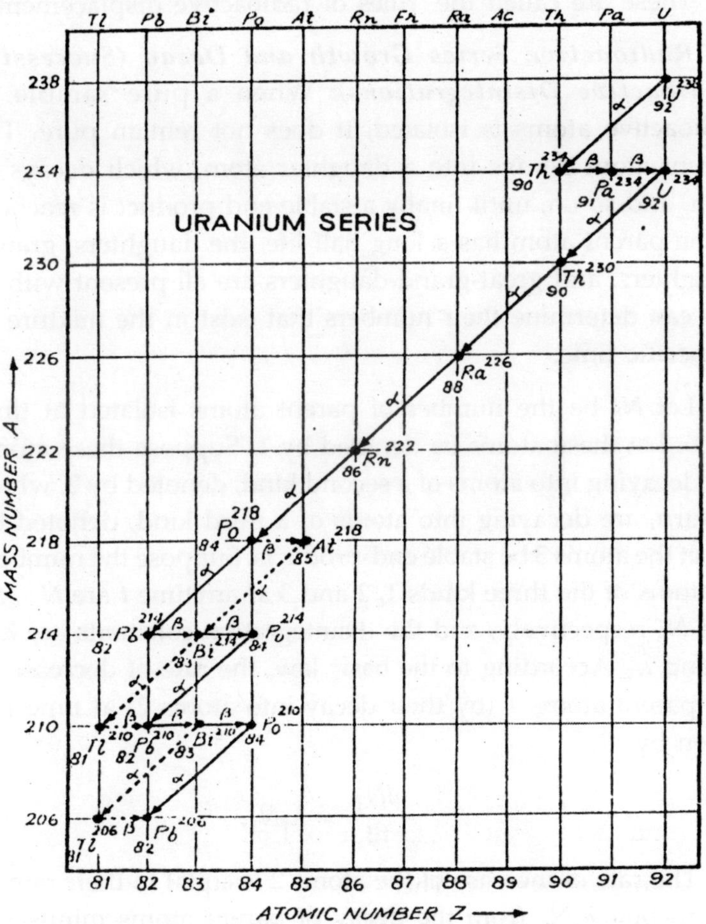

Laws of Radioactive Displacement

(i) When a radioactive atom emits an α-particle, the product atom shifts in the periodic table two steps in the direction of lower atomic number and its mass number is lowered by 4 units.

(ii) On the emission of a β-particle, the product atom shifts one step in the direction of increasing atomic number.

These are called the 'rules of radioactive displacement'.

Radioactive Series Growth and Decay (Successive Radioactive Disintegrations): When a pure sample of radioactive atoms is isolated, it does not remain pure. The parent atom decays into a daughter atom, which decays in turn, and so on, until finally a stable end-product is reached. If the parent atom has a long half-life, the daughters, grand-daughters, and great-grand-daughters are all present with it. We can determine their numbers that exist in the mixture at a specific time.

Let N_0 be the number of parent atoms isolated at time $t = 0$. Let these atoms be denoted by 1. Suppose these atoms are decaying into atoms of a second kind, denoted by 2, which in turn, are decaying into atoms of a third kind, denoted by 3. Let the atoms 3 be stable end-products. Suppose the numbers of atoms of the three kinds 1, 2 and 3 at anytime t are N_1, N_2 and N_3 respectively, and the disintegration constants are λ_1, λ_2 and λ_3. According to the basic law, the rate of decrease of the parent atoms 1 (by their decay into atoms 2) at time t is given by

$$-\frac{dN_1}{dt} = \lambda_1 N_1. \qquad \text{... } (i)$$

The rate of increase of the atoms 2 is equal to their rate of production $\lambda_1 N_1$ from the decay of parent atoms minus the rate of their own decay $\lambda_2 N_2$ into the atoms 3. Thus

$$\frac{dN_2}{dt} = \lambda_1 N_1 - \lambda_2 N_2. \qquad (ii)$$

The rate of increase of the (stable) atoms 3 is equal to the rate of decay of the atoms 2 into the atoms 3. Thus

$$\frac{dN_3}{dt} = \lambda_2 N_2. \qquad (iii)$$

From Rutherford-Soddy equation, the number N_1 of parent atoms 1 at the time t can be directly written as

$$N_1 = N_0 e^{-\lambda_1 t}. \qquad \dots (iv)$$

Substituting this value of N_1 in eq. (*ii*), we get

$$\frac{dN_2}{dt} = \lambda_1 N_0 e^{-\lambda_1 t} - \lambda_2 N_2$$

or
$$\frac{dN_2}{dt} + \lambda_2 N_2 = \lambda_1 N_0 e^{-\lambda_1 t}$$

Multiplying both sides by $e^{-\lambda_2 t}$, we obtain

$$e^{\lambda_2 t}\frac{dN_2}{dt} + \lambda_2 N_2 e^{\lambda_2 t} = \lambda_1 N_0 e^{(\lambda_2 - \lambda_1)t}$$

or
$$\frac{d}{dt}\left(N_2 e^{\lambda_2 t}\right) = \lambda_1 N_0 e^{(\lambda_2 - \lambda_1)t}$$

Integrating :
$$\lambda_2 e^{\lambda_2 t} = \frac{\lambda_1}{\lambda_2 - \lambda_1} N_o e^{(\lambda_2 - \lambda_1)t} + C,$$

where C is a constant of integration.

At $t = 0$, $N_2 = 0$ (because only parent atoms 1 were present initially).

$$\therefore \qquad C = -\frac{\lambda_1}{\lambda_2 - \lambda_1} N_o.$$

This gives

$$N_2 e^{\lambda_2 t} = \frac{\lambda_1}{\lambda_2 - \lambda_1} N_o e^{(\lambda_2 - \lambda_1)t} - \frac{\lambda_1}{\lambda_2 - \lambda_1} N_0.$$

or $$N2 = \frac{\lambda_1}{\lambda_2 - \lambda_1} N_0 \left(e^{-\lambda_1 t} - e^{-\lambda_2 t}\right). \quad \text{... } (v)$$

On substituting this value of N_2 in eq. (*iii*) and then integrating and applying the condition that at $t = 0, N_3 = 0$ we get

$$N_3 = N_0\left(1 + \frac{\lambda_1}{\lambda_2 - \lambda_1} e^{-\lambda_2 t} - \frac{\lambda_2}{\lambda_2 - \lambda_1} e^{-\lambda_1 t}\right). \quad \text{... } (vi)$$

Eq. (*iv*), (*v*) and (*vi*) represent the numbers of atoms 1, 2 and 3 respectively present in the mixture at a specified time *t*.

Fig. shows the numbers of atoms 1, 2 and 3 as a function of time when it is assumed that atoms 2 have a half-life longer than atoms 1, while atoms 3 are stable. Initially there are N_0 parent atoms. The number N_1 of the parent atoms l decreases exponentially according to eq. (*iv*). The number N_2 is initially zero, it increases, passes through a maximum, and then decreases gradually in accordance to eq. (*v*). The number of atoms N_3 of the stable end-product increases steadily with time, eventually approaching N_0.

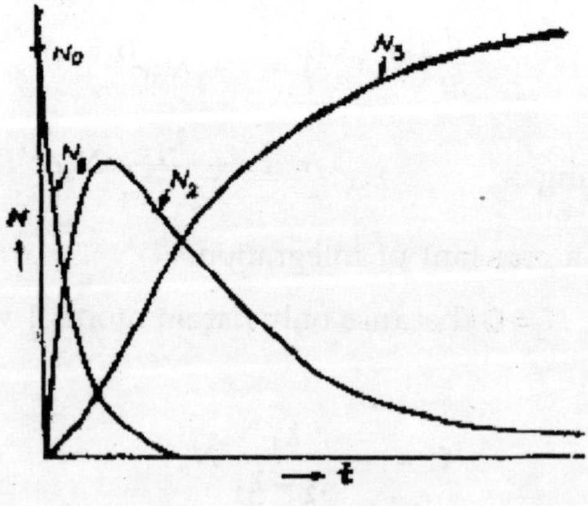

Transient Radioactive Equilibrium

In a, radioactive series in which the parent has a very long half-life, a state is reached after a fairly long time when each daughter-product is formed at the same rate as it decays. When this condition is reached, the proportions of the different radioactive atoms in the mixture do not change with time, and the parent is said to be in 'secular radioactive equilibrium' with its daughter-products. We can prove it from the theory of successive disintegrations :

Suppose that the parent atoms 1 have a very long half-life (*i.e.* decay much more slowly) than any of the decay products ($T_1 >> T_2$). Then

$$\lambda_1 \approx 0 \text{ , and } \lambda_1 << \lambda_2, \lambda_1 << \lambda_3 \text{ , and so on.}$$

Substituting $\lambda_1 \approx 0$ in eq. (iv), we get

$$N_1 \approx N_0. \qquad [\therefore e^{-\lambda_1 t} \approx 1 \text{ when } \lambda_1 \approx 0]$$

Substituting $\lambda_1 << \lambda_2$ in eq. (v); we get

$$N_2 = \frac{\lambda_1}{\lambda_2} N_0 \left(1 - e^{-\lambda_2 t}\right) \qquad \left[\therefore \frac{\lambda_1}{\lambda_2 - \lambda_1} \approx \frac{\lambda_1}{\lambda_2}\right]$$

$$= \frac{\lambda_1}{\lambda_2} N_1 \left(1 - e^{-\lambda_2 t}\right) \qquad [\therefore N_1 \approx N_0]$$

After a sufficient period of time *i.e.* when t becomes sufficiently large, $e^{-\lambda_2 T}$ becomes negligibly small, so that

$$N_2 = \frac{\lambda_1}{\lambda_2} N_1$$

or

$$\lambda_1 N_1 = \lambda_2 N_2.$$

$\lambda_1 N_1$ is the rate of production of the atoms 2 due to the decay of the parent atoms *l*, and $\lambda_2 N_2$ is their own rate of decay into the atoms 3, and these two rates have become equal.

Thus after a sufficient time, the activities of parent and daughter become equal. The condition is known as 'secular equilibrium' and is shown in Fig.

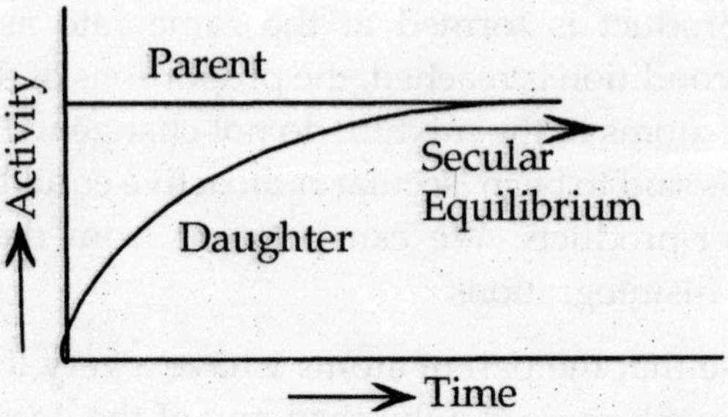

For a series of several daughter-products the above equation can be expanded as

$$\lambda_1 N_1 = \lambda_2 N_2 = \lambda_3 N_3 = \dots \dots$$

or in term of half-lives ($T = \log_e 2/\lambda$) ;

$$\frac{N_1}{T_1} = \frac{N_2}{T_2} = \frac{N_3}{T_3} = \dots \dots$$

The secular equilibrium has been established in the uranium series, the (parent) uranium having a half-life (4.5×10^9 years) so long that it takes 50×10^6 years for its quantity in a rock to change by 1%.

A different type of equilibrium exists when the parent is longer-lived than the daughter ($\lambda_1 < \lambda_2$), but the half-life of the parent is not very long ($T_1 \simeq T_2$). In this case, λ_1 cannot be assumed zero. After t becomes sufficiently large, $e^{-\lambda_2 t}$ becomes negligible compared with $e^{-\lambda_1 t}$ so that the number of the daughter atoms is given by (see eq. v)

$$N_2 = \frac{\lambda_1}{\lambda_2 - \lambda_1} N_0 e^{-\lambda_1 t}$$

Comparing it with eq. (iv) and remembering that $\lambda_1 < \lambda_2$, we find that the daughter eventually decays with the same half-life as the parent. Since $N_0\, e^{-\lambda_1 t} = N_1$, we have

$$N_2 = \frac{\lambda_1}{\lambda_2 - \lambda_1} N_1$$

or $$\frac{N_1}{N_2} = \frac{\lambda_2 - \lambda_1}{\lambda_1}.$$

Thus after a sufficient time the ratio of parent atoms to daughter atoms becomes constant, and both eventually decay. This condition is called 'transient equilibrium', and is shown in Fig.

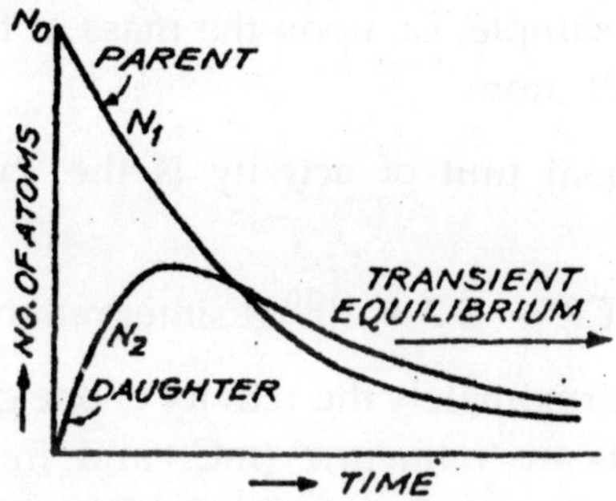

When the parent has a shorter half-life time than the daughter ($T_1 < T_2$ or $\lambda_1 > \lambda_2$), no state of equilibrium is attained. If initially we have only the parent atoms, then as the parent atoms decay, the daughter atoms increase in number, pass through a maximum, and eventually decay with their own half-life.

Activity of Radioactive Substances : The activity of a sample of any radioactive material is the rate at which its constituent atoms disintegrate. Thus if dN be the number of atoms which disintegrate during a time-interval dt, the activity R of the sample is given by

$$R = -\frac{dN}{dt}$$

The negative sign indicates that the number of atoms is decreasing with time.

If at any time the number of atoms in the sample is N, then it is an experimental fact that the rate of disintegration of the atoms is proportional to N, that is,

$$-\frac{dN}{dt} \propto N = \lambda N,$$

where λ is the decay constant for the particular atom (rather the particular isotope). Thus

$$R = \lambda N.$$

Thus the activity of the sample depends upon the number of atoms in the sample, *i.e.* upon the mass of the sample and upon the type of atom.

The traditional unit of activity is the 'curie' (*Ci*). It is defined as:

$$1 \text{ curie } (Ci) = 3.7 \times 10^{10} \text{ disintegrations/sec.}$$

1 curie is approximately the activity of one gram of radium. The smaller units are 'millicurie' (mCi) and 'microcurie' (μCi).

$$1 \text{ m}Ci = 10^{-3}\, Ci = 3.7 \times 10^{7} \text{ disintegrations/sec.}$$

$$1\ \mu Ci = 10^{-6}\, Ci = 3.7 \times 10^{4} \text{ disintegrations/sec.}$$

Another unit of activity is 'rutherford', which is by definition

$$1 \text{ rutherford} = 10^{6} \text{ disintegrations/sec.}$$

The SI unit of activity is the 'becquerel' (Bq); named after the discoverer of radioactivity :

$$1 \text{ becquerel (Bq)} = 1 \text{ disintegration/sec. so that}$$

$$1\ Ci = 3.7 \times 10^{10} \text{ Bq.}$$

***Determination of Half-life* (*T*)** : The half-lives of the known radioactive substances cover an enormous range from about more than 10^{15} years to about less than 10^{-6} second. Although many of these can be determined by direct measurements in the laboratory ; those which are extremely long or short are estimated by less direct methods.

The moderate half-lives can be determined by measuring the activity of the given sample for a reasonable time-interval.

We know that the activity of a sample at an instant t is given by

$$R = -\frac{dN}{dt} = \lambda N,$$

where N is the number of radioactive atoms in the sample at time t, and λ is the disintegration constant. From Rutherford-Soddy law,

$$N = N_0 e^{-\lambda t},$$

where N_0 is the number of atoms in the sample at time $t = 0$.

$$\therefore \quad R = \lambda N_0 e^{-\lambda t},$$

But λN_0 is the initial activity R_0 at time $t = 0$. Therefore

$$R = R_0 e^{-\lambda t}.$$

On taking logarithms :

$$\log_e R = \log_e R_0 - \lambda t$$

$$\text{or} \quad 2.3026 \log_{10} R = 2.3026 \log_{10} R_0 - \lambda t$$

$$\text{or} \quad \log_{10} R = \log_{10} R_0 - 0.434\,\lambda t.$$

Hence, when the logarithm of the measured activity R is plotted against the time t, a straight line is obtained whose slope is equal to $- 0.434\,\lambda$. Thus, by measuring the slope, the disintegration constant λ and hence the half-life T ($= 0.693/\lambda$) can be calculated.

The activity (number of atoms disintegrating per second) is measured by a Geiger counter or a scintillation counter which counts the particles emitted per second from the radioactive sample, and each emitted particle represents the disintegration of one atom.

For determining very long lives, the activity R in a given number of radioactive atoms N is measured in specially-designed counting systems and the following relation is used:

$$R = \lambda N$$

so that $$\lambda = \frac{R}{N}.$$

N is determined by finding the total mass M of the radioactive substance $\left(N = \frac{M}{\text{mass number}} \times \text{avogadro's number}\right)$.

The half-life of a very short-lived radioactive atom can be obtained, if it emits α-particles, by measuring the range of these particles and applying Geiger-Nuttall rule.

Radioactive Dating (Age of Earth): The decay of radioactive elements and its complete independence from physical and chemical conditions gives us a method for estimating the ages of old rocks and earth's crust. If a radioactive element and its decay products remain trapped in a rock (since it first solidified), the measurement of the proportion of the end-product to the parent element enables us to determine the age of the rock.

It is believed that the element U^{288} was formed at some time in the past when the earth was solidified and has been remained trapped in the rocks and earth's crust. Since then it has been decaying forming finally a stable isotope of lead (Pb^{206}), and has reached a state of secular equilibrium. Now, the disintegration constant of U^{288} is 1.54×10^{-10} year^{-1}. It means that, 1 gm of U^{238} produces an amount of Pb^{206} equal to

$$\frac{206}{238} \times (1.54 \times 10^{-10}) = 1.33 \times 10^{-10} \text{ gm/year}.$$

Hence, if (Pb^{206}/U^{238}) be the mass of Pb^{206} per gram of U^{233} in a specimen of earth's crust, then the time T that has elapsed since the disintegration of U^{238} started is given by (neglecting the decrease in the amount of U^{238})

$$T = \frac{\left(Pb^{206}/U^{238}\right)}{1.33 \times 10^{-10}} = (Pb^{206}/U^{288}) \times (7.5 \times 10^{9}) \text{ years}.$$

This can be taken to be the age of the earth. Measurements on the oldest rocks has established the age of the earth to be $\simeq 4 \times 10^9$ years.

PROBLEMS

1. ***The isotope $_{92}U^{238}$ successively undergoes eight alpha decays and six beta decays. What is the resulting isotope ?***

Solution : The isotope $_{92}U^{238}$ has mass number 238 and atomic number 92. An α-decay reduces the mass number by 4 and atomic number by 2. A β-decay simply increases the atomic number by 1. Hence the new mass number is 238— (8 × 4) = 206, while the new atomic number is 92 – (8 × 2) + (6 × 1) = 82. The new isotope is $_{82}Pb^{206}$.

2. ***The thorium decay series begins with $_{90}Th^{232}$ and ends with $_{82}Pb^{208}$. How many α- and how many β-decays occur in the series ?***

Solution : An α-decay reduces the mass number by 4 and atomic number by 2. A β-decay simply increases the atomic number by 1. In the series $_{90}Th^{232} \rightarrow {_{82}Pb^{208}}$, the mass number is reduced by (232—208) = 24. Hence there occur 6 α-decays.

6 α-decays reduce the atomic number by 12. But, in the series, the atomic number is reduced by only (90 — 82) = 8, *i.e.*, there is also an increase of 4 in the atomic number. Hence there occur 4 β-decays.

3. ***Calculate the radioactive constant for an element whose half-life period is 20 years.***

Solution : The radioactive constant λ and the half-life period *T* of a radioactive element are related as

$$\lambda = \frac{\log_e 2}{T} = \frac{0.693}{T}$$

Here T = 20 years.

$$\therefore \qquad \lambda = \frac{0.693}{20 \text{ years}} = 0.03465 \text{ per year.}$$

4. *Find the half-life time and mean-life time of a radioactive substance of which the decay constant is 4.28 × 10^{-4} per year.*

Solution : The half life is

$$T = \frac{0.693}{\lambda} = \frac{0.693}{4.28 \times 10^{-4} \text{ / year}}$$
$$= 1619 \text{ years.}$$

The mean-life is

$$\overline{T} = \frac{1}{\lambda} = \frac{1}{4.28 \times 10^{-4} \text{ / year}} = 2336 \text{ years.}$$

5. *Ten milligrams of a radioactive substance of half-life period two years is kept in store for four years. How much of the substance remains unchanged ?*

Solution : The time-interval in which the mass (or number of atoms) of a radioactive element decays to one half of its initial value, is called the 'half-life' of the element. Thus, if the initial quantity of a radioactive substance be N_0 , then the quantity N of the substance left after n half-lives is given by

$$N = N_0 \left(\frac{1}{2}\right)^n .$$

Here $\qquad N_0 = 0 \text{ mgm and } n = \frac{4 \text{ years}}{2 \text{ years}}.$

Therefore the mass of the substance remaining after 2 half-lives is

$$N = 10 \text{ mgm} \left(\frac{1}{2}\right)^2 = 2.5 \text{ mgm.}$$

6. *The half-life of a radioactive nucleus is 2.5 days. What % of the original substance will have disintegrated in 7.5 days?*

Solution : If N_0 be the initial quantity of a radioactive substance, then the quantity N left over after n half-lives is given by

Here $$n = \frac{7.5 \text{ days}}{2.5 \text{ days}} = 3$$

$$\therefore \quad \frac{N}{N_0} = \left(\frac{1}{2}\right)^3 = \frac{1}{8}.$$

The fraction disintegrated is $1 - \frac{1}{8} = \frac{7}{8}$, that is, 87.5%.

7. The half-life of radon is 3.82 days. What fraction of a freshly separated sample of this nuclide will disintegrate in one day ? If the sample contains initially 2.71 x 10^{15} atoms, how many atoms will disintegrate during the first day ?

Solution : If N_0 be the number of atoms at $t = 0$, and N the number at time t, then by Rutherford-Soddy law

$$N = N_0 e^{-\lambda t}$$

or $$\log_e \frac{N_0}{N} = \lambda t.$$

Here $$\lambda = \frac{0.693}{T} = \frac{0.693}{3.82} \text{ per day and } t = 1 \text{ day}.$$

$$\therefore \quad 2.3026 \log_{10} \frac{N_0}{N} = \frac{0.693 \times 1}{3.82}$$

or $$\log_{10} \frac{N_0}{N} = \frac{0.693}{3.82 \times 2.3026} = 0.0788$$

or $$\frac{N}{N_0} = \text{antilog } (0.0788) = 1.199$$

or fraction disintegrated,

$$1 - \frac{N}{N_0} = 1 - \frac{1}{1.199} = \frac{0.199}{1.199}$$

$$= 0.166 = 16.6\%.$$

If $N_0 = 2.71 \times 10^{15}$ atoms, then the number of disintegrated atoms during the first day is

$$N_0 - N = N_0 \left(1 - \frac{N}{N_0}\right)$$

$$= 2.71 \times 10^{15} \times 0.166$$

$$= 4.5 \times 10^{14}.$$

8. *The half-life of Na ^{24} is 15 hr. How long does it take for 87.5% of this isotope to decay?*

Or

The half-life of a radioactive substance is 15 hr. Calculate the period in which 12.5% of the initial quantity of the substance will be left over.

Solution : Let t be the time period in which 87.5% of the initial quantity of the radioactive substance is decayed, or 12.5% is left over. Now,

$$\text{half-life, } T = 15hr, .$$

$$\therefore \text{ disintegration constant, } \lambda = \frac{0.693}{15} = 0.0462 \text{ per hr.}$$

If N_0 be the number of atoms at $t = 0$, then the number N at time t is given by (Rutherford-Soddy law)

$$N = N_0 \, e^{-\lambda t}$$

Here $$\frac{N}{N_0} = 12.5\% = \frac{12.5}{100} = \frac{1}{8}.$$

$$\therefore \quad \frac{1}{8} = e^{-\lambda t}$$

$$e^{\lambda t} = 8.$$

or $$\lambda t = \log_e 8 = 2.3026 \log_{10} 8$$

$$\therefore \qquad t = \frac{2.3026 \log_{10} 8}{\lambda}$$

$$= \frac{2.3026 \times 0.9031}{0.0462} = 45 \text{ hr.}$$

9. ***The half-life of $_{11}Na^{24}$ is 15 hours. How long will it take for 93.75% of a sample of this isotope to decay ?***

[Ans. 60 hr.**]**

10. ***The half-life of radon is 3.8 days. After how many days will only one-twentieth (1/20) of a radon sample be left over?***

Hint : $\frac{N}{N_0} = \frac{1}{20}$. **[Ans.** 16.5 days.**]**

11 ***The half-life of radium is 1590 years. In how many years will 1 gm of radium (i) be reduced to 1 centigram, (ii) loses 1 centigram.***

Solution : (*i*) Let t be the time in which 1 gm of radium will be reduced to 1 centigram ($\frac{1}{100}$ =0.01 gm) . The decay constant is

$$\lambda = \frac{0.693}{T} = \frac{0.693}{1590} \text{ per year.}$$

If N_0 be the number of atoms at $t = 0$, and N the number at time t then, by Rutherford-Soddy law

$$N = N_0 e^{-\lambda t}$$

Here $$\frac{N}{N_0} = \frac{1 \text{ centigram}}{1\text{gm}} = 0.01 = \frac{1}{100}.$$

$$\therefore \qquad \frac{1}{100} = e^{-\lambda t}$$

or $$e^{\lambda t} = 100$$

or $$\lambda t = \log_e 100 = 2.3026 \times \log_{10} 100$$

$$= 2.3026 \times 2$$

$$\therefore \qquad t = \frac{2.3026 \times 2}{\lambda}$$

$$= \frac{2.3026 \times 2}{0.693/1590} = 10566 \text{ years.}$$

(*ii*) In this case in the beginning (at t=0) the radium is 1 gm. In a time t (say), it loses 1 centigram (= 0.01 gm), that is, it remains (1–0.01) = 0.99 gm.

Thus
$$\frac{N}{N_0} = \frac{0.99}{1} = \frac{99}{100}.$$

$$\therefore \qquad \frac{99}{100} = e^{-\lambda t}$$

or
$$e^{\lambda t} = \frac{100}{99}$$

or
$$\lambda t = \log_e \frac{100}{99} = 2.3026 \log_{10} \frac{100}{99}$$

$$= 2.3026 \times 0.0044.$$

$$\therefore \qquad t = \frac{2.3026 \times 0.0044}{\lambda}$$

$$= \frac{2.3026 \times 0.0044}{0.693/1590} = 23.2 \text{ years.}$$

12. ***10 gram of a radioactive substance is reduced by 2.5 mg in 6 years through α-decay. Evaluate half-life time and mean-life time of the substance.***

Solution: 10 gm of the substance is reduced to (10 – 0.0025) = 9.9975 gm in 6 years. .

Let N_0 be the amount of the substance at t = 0 and N at time t. Then, we have

$$\frac{N}{N_0} = e^{-\lambda t},$$

where λ is decay constant. Here, $\frac{N}{N_0} = \frac{9.9975}{10}$ and $t = 6$ years.

$$\therefore \qquad \frac{9.9975}{10} = e^{-6\lambda}$$

$$\text{or} \qquad e^{6\lambda} = \frac{10}{9.9975}$$

$$\text{or } \lambda = \frac{1}{6} \times 2.3026 \times (\log_{10} 10 - \log_{10} 9.9975)$$

$$= \frac{1}{6} \times 2.3026 \times (1.0000 - 0.9999)$$

$$= 3.84 \times 10^{-5} \text{ per yr.}$$

$$\therefore \text{ half-life, } T = \frac{0.693}{\lambda} = \frac{0.693}{3.84 \times 10^{-5}} = 1.8 \times 10 \text{ years}$$

$$\text{and mean-life, } \bar{T} = \frac{1}{\lambda} = \frac{1}{3.84 \times 10^{-5}} = 2.6 \times 10^4 \text{ years.}$$

13. ***A radioactive element disintegrates for an interval of time equal to its mean life. (i) What fraction of element remains? (ii) What fraction has disintegrated ?***

Hint: $\bar{T} = 1/\lambda$.

[Ans. (*i*) $1/e$, (*ii*) $(e-1)/e$**]**

14. ***$_{83}Bi^{212}$ decays to $_{81}Tl^{208}$ by α-emission in 34% of the disintegrations and to $_{84}Po^{212}$ by β-emission in 66% of the disintegrations. If the total half-value period is 60 5 minutes, find the decay constants for α and β emissions.***

Solution : Certain nuclei break up in two different ways, either by α-emission or by β-emission, giving rise to two different product nuclei. The probability of disintegration is the sum of separate probabilities and

$$\lambda = \lambda_\alpha + \lambda_\beta.$$

The half-life is $T = \dfrac{0.693}{\lambda} = \dfrac{0.693}{\lambda_\alpha + \lambda_\beta}$. Thus, here

$$\lambda_\alpha + \lambda_\beta = \frac{0.693}{T} = \frac{0.693}{60.5 \text{ min}} = 0.01145 \text{ min}^{-1}.$$

Since $\quad -\dfrac{dN}{dt} = \lambda N$, we can write

$$0.34 = \lambda_\alpha N$$

and $\quad 0.66 = \lambda_\beta N.$

These give $\quad \dfrac{0.34}{0.66} = \dfrac{\lambda_\alpha}{0.01145 - \lambda_\alpha}.[\therefore \lambda_\alpha + \lambda_\beta = 0.01145]$

Solving $\quad \lambda_\alpha = 0.003893 \text{ min}^{-1}$

and $\quad \lambda_\beta = 0.007557 \text{ min.}^{-1}$

15. ***The mean lives of a radioactive substance are 1620 and 405 years for α-emission and β-emission respectively. Find out the time during which three-fourth of a sample will decay if it is decaying both by α-emission and β-emission simultaneously.***

Solution: The decay constant for a- and β-emission are $\dfrac{1}{1620}$ and $\dfrac{1}{405}$ per year respectively.

$\therefore$ total decay constant, $\lambda = \dfrac{1}{1620} + \dfrac{1}{405} = \dfrac{1}{324}$ per year.

If N_0 be the amount of a sample at $t = 0$ and N the amount left over at time t, then by the decay law

$$N = N_o e^{-\lambda t}$$

After $\dfrac{3}{4}$th part has been disintegrated, we have $N = N_0/4$.

Therefore

$$\frac{1}{4} = e^{-\lambda t}$$

or $$e^{\lambda t} = 4$$

or $$t = \frac{1}{\lambda} \log_e 4 = \frac{1}{\lambda} (2.3026 \log_{10} 4)$$

$$= 324 (2.3026 \times 0.6020) = 449 \text{ years.}$$

16. *The activity of a radioactive substance decreases to 1/64 of its original value in 21 years. Calculate the half-life of the substance.*

Solution: The activity (rate of disintegration) at any instant is directly proportional to the number of radioactive atoms present at that instant. Thus a decrease in activity means a corresponding decrease in the number of atoms. If N_0 be the number of atoms initially and N the number left after a time-interval t (21 years), then

$$\frac{N}{N_0} = \frac{1}{64}.$$

By Rutherford-Soddy law, we have

$$\frac{N}{N_0} = e^{-\lambda t} = \frac{1}{64}$$

or $$e^{\lambda t} = 64$$

or $$\lambda t = \log_e 64 = 2.3026 \times \log_{10} 64.$$

Here $$t = 21 \text{ years.}$$

$$\therefore \quad \lambda = \frac{0.3026 \times 1.8062}{21} = 0.198 \text{ per year.}$$

Hence the half-life is

$$T = \frac{0.693}{\lambda} = \frac{0.693}{0.198} = 3.5 \text{ years}$$

17. *A radioactive substance at a given instant emits 4750 particles per minute. Five minutes later it emits 2700 particles*

per minute. Find the decay constant and half-life of the substance.

Solution : The rate of disintegration of a substance at any instant is proportional to the number of atoms at that instant. Thus, if N_0 be the number of atoms at $t = 0$, when the rate of disintegration is 4750 min^{-1}, and N be the number of atoms at $t = 5$ min when the rate of disintegration is 2700 min^{-1}, then

$$\frac{N}{N_0} = \frac{2700}{4750}.$$

From Rutherford-Soddy law, $N/N_0 = e^{-\lambda t}$

$\therefore\ \lambda t = \log_e (N/N_0) = 2.3026 \log_{10} (N_0/N)$.

Here t=5 min and $N_0/N = 4750/2700 = 1.76$.

$$\therefore \qquad \lambda = \frac{2.3026 \log_{10} 1.76}{5 \text{ min}}$$

$$= \frac{2.306 \times 0.2455}{5 \text{ min}} = 0.113 \text{ min}^{-1}.$$

The half-life T of the material is

$$T = \frac{0.693}{\lambda} = \frac{0.693}{0.113} = 6.13 \text{ min}.$$

18. ***The activity of a radioactive substance is reduced to half of its original value in 3.86 days. Find the time-interval in which its activity will become 1%. What is the average life of the atom of the substance ?***

Solution : The activity is proportional to the number of atoms. It is reduced to half in 3.86 days. It means the half-life T is 3.86 days.

$$\therefore \qquad \lambda = \frac{0.693}{3.86} = 0.1795 \text{ per day}.$$

Let t be the time in which the number of atoms reduces to 1%. By the decay law,

$$= \frac{N}{N_0} = e^{-\lambda t}$$

Here $\frac{N}{N_0} = 1\% = \frac{1}{100}.$

$\therefore$ $\frac{1}{100} = e^{-\lambda t}$

or $e^{\lambda t} = 100$

or $t = \frac{1}{\lambda} \log_e 100 = \frac{1}{\lambda} (2.3026 \log_{10} 100).$

$$= \frac{2.3026 \times 2}{0.1795} = 25.6 \text{ days.}$$

The average life of the atom is

$$\bar{T} = \frac{1}{\lambda} = \frac{1}{0.1795} = 5.57 \text{ days.}$$

19. ***The half-life of $_{92}U^{238}$ against α-decay is 4.5×10^9 years. Find the activity of 1 gm of $_{92}U^{238}$. (Avogadro Number = 6.02 $\times 10^{28}$ per gin-atom).***

Solution: 1 gm-atom of U^{238} has a mass of 238 gm and contains 6.02×10^{23} atoms (the Avogadro's number). Therefore, the number of atoms in 1 gm of U^{238} is

$$N = \frac{6.02 \times 10^{23}}{238} = 2.53 \times 10^{21}.$$

Its half-life is $T = 4.5 \times 10^9$ years. Therefore its decay constant is

$$\lambda = \frac{0.693}{T} = \frac{0.693}{4.5 \times 10^9}$$

$$= 1.54 \times 10^{-10} \text{ year}^{-1}.$$

Now, the activity R, or the rate of disintegration $-\frac{dN}{dt}$ of a substance at any instant is proportional to the number of atoms N present at that instant *i.e.*

$$R = -\frac{dN}{dt} = \lambda N$$

$$= (1.54 \times 10^{-10}) \times (2.53 \times 10^{21})$$

$$= 3.9 \times 10^{11} \text{ year}^{-1}$$

$$= \frac{3.9 \times 10^{11}}{365 \times 24 \times 60 \times 60}$$

$$= 1.23 \times 10^{4} \text{ integrations/sec}$$

$$= 1.23 \times 10^{4} \text{ Bq.}$$

20. ***Calculate the radioactivity in curie for 1 gm of $_{38}Sr^{90}$ with half-life period of 28 years. Given : Avogadro number =6.025 × 10^{26} molecules/kg molecule.***

Solution : The intensity of radioactivity for 1 gm of substance is given by

$$R = \lambda N,$$

where λ is decay constant and N is the number of atoms in 1 gm.

Here $$\lambda = \frac{0.693}{\text{half} - \text{life}}$$

$$= \frac{0.693}{28 \text{ years}} = \frac{0.693}{28 \times 365 \times 24 \times 60 \times 60 \text{ sec}}$$

$$= 7.85 \times 10^{-10} \text{ sec}^{-1}.$$

and $$N = \frac{6.025 \times 10^{23} \text{ atoms / gm - atom}}{90 \text{ gm / gm} - \text{atom}}$$

$$= 6.69 \times 10^{21} \text{ atoms/gm.}$$

$\therefore$ $$R = (7.85 \times 10\text{-}^{10} \text{ sec}^{-1})\,(6.69 \times 10^{21} \text{ atoms})$$

$$= 5.25 \times 10^{12} \text{ disintegrations/sec.}$$

Now, 1 curie = 3.7×10^{10} disintegration/sec.

$$\therefore \quad R = \frac{5.25 \times 10^{12}}{3.7 \times 10^{10}} = 142 \text{ Ci.}$$

21. *One gm of Ra^{226} has an activity of 1 curie. Determine the half-life of radium. Avogadro number is 6.02 × 10^{28}.*

Solution : 1 gm-atom of radium has a mass of 226 gm, and contains 6.02×10^{23} atoms (the Avogadro's number). Therefore, the number of atoms in 1 gram of radium is

$$N = \frac{6.02 \times 10^{23}}{225} = 2.66 \times 10^{21}.$$

Now, the activity R of a radioactive sample is the rate of disintegration of its atoms;

$$R = -\frac{dN}{dt} = \lambda N,$$

where λ is the disintegration constant.

Here R = 1 curie = 3.7×10^{10} disintegrations/sec and $N = 2.66 \times 10^{21}$.

$$\therefore \quad \lambda = \frac{R}{N} = \frac{3.7 \times 10^{10}}{2.66 \times 10^{21}} = 1.39 \times 10^{-11} \text{ sec}^{-1}$$

The half-life is

$$T = \frac{0.693}{\lambda} = \frac{0.693}{1.39 \times 10^{-11}} = 4.94 \times 10^{10} \text{ sec.}$$

Now, 1 year = $365 \times 24 \times 60 \times 60 = 3.15 \times 10^{7}$ sec.

$$\therefore \quad T = \frac{4.98 \times 10^{10}}{3.15 \times 10^{7}} = 1581 \text{ years}$$

The average-life is

$$\overline{T} = \frac{1}{\lambda} = \frac{1}{1.39 \times 10^{-11}} = 7.19 \times 10^{10} \text{ sec}$$

$$= \frac{7.19 \times 10^{10}}{3.15 \times 10^{7}} = 2282 \text{ years.}$$

22. A sample of U^{234} for which the decay constant is 8.78 × 10^{-14} per sec undergoes 3.7 × 10^8 disintegrations per second. Find the mass of the sample. Avogadro number = 6.03 × 10^{28}.

Solution: The activity in N atoms is

$$R = \lambda N.$$

Here R = 3.7 × 10^8 disintegrations/sec and

$$\lambda = 8.78 \times 10^{-14} \text{ per sec.}$$

$$\therefore \quad N = \frac{R}{\lambda} = \frac{3.7 \times 10^8}{8.78 \times 10^{-14}} = 4.21 \times 10^{21} \text{ atoms.}$$

Thus there are 4.21 × 10^{21} atoms in the given sample of U^{234}.

Mass of 6.03 × 10^{23} atoms of U^{234} = 234 gm.

∴ mass of 4.21 × 10^{21} atoms of

$$U^{234} = \frac{234 \times \left(4.21 \times 10^{21}\right)}{6.03 \times 10^{23}} = 1.63 \text{ gm.}$$

23. The half-life of a cobalt radio-isotope is 5.3 years. What strength will a milli-curie source of the isotope have after a period of one year and 5.3 years?

Solution: The strength, (activity) R of a radioactive source containing N atoms is given by

$$R = \lambda N,$$

where λ is decay constant.

$$\text{Here,} \qquad \lambda = \frac{0.693}{5.3 \text{ yr}} = 0.131 \text{ per yr.}$$

Let N_0 be the number of atoms at $t = 0$ when the strength is 1 milli-curie, and N the number at a time $t = 1$ year when the strength is R' (say). Then

$$1 = \lambda N_0 \text{ and } R' = \lambda N.$$

$$\therefore \qquad \frac{N}{N_0} = R'.$$

But from Rutherford-Soddy law, $\dfrac{N}{N_0} = e^{-\lambda t}$

$$\therefore \qquad R' = e^{-\lambda t}$$

or $$\log_e\left(\frac{1}{R'}\right) = \lambda t.$$

Here λ = 0.131 per year and t = 1yr.

$$\therefore \qquad \log_e\left(\frac{1}{R'}\right) = 0.131 \times 1.$$

or $$\log_e\left(\frac{1}{R'}\right) = \frac{0.131 \times 1}{2.3026} = 0.0569$$

or $$\frac{1}{R'} = \text{antilog } 0.0569 = 1.14$$

$$\therefore \qquad R' = \frac{1}{1.14} = 0.88 \text{ milli-curie.}$$

The strength after 5.3 years (half-life) will become one-half, *i.e.*, 0.5 milli-curie.

24. *The isotopes U^{238} and U^{235} occur in nature in the ratio 140 : 1. Assuming that at the time of earth's formation they were present in equal ratio, make an estimation of the age of the earth. Th e half-lives of U^{238} and U^{235} are 4.5 × 10^9 years and 7.13 × 10^8 years respectively. ($\log_{10}$ 140 = 2.1461, $\log_{10}$ 2 = 0.3010)*

Solution : Let N_1 and N_2 be the number of atoms of U^{238} and U^{235} , and T_1 and T_2 their half-lives.

From the decay formula $N = N_0^{e - \lambda t}$, we have

$$\frac{N_1}{N_2} = e^{(\lambda_2 - \lambda_1)t},$$

where t is the elasped time. From this, we have

$$t = \frac{\log_e (N_1/N_2)}{\lambda_2 - \lambda_1}$$

$$= \frac{\log_e (140/1)}{\log_e 2\left(\frac{1}{T_2} - \frac{1}{T_1}\right)} \qquad \left[\therefore \lambda = \frac{\log_e 2}{T}\right]$$

$$= \frac{\log_{10} 140}{\log_{10} 2}\left(\frac{T_2 T_1}{T_1 - T_2}\right)$$

$$= \frac{2.1461}{0.3010} \times \frac{7.13 \times 10^8 \times 4.5 \times 10^9}{37.87 \times 10^8}$$

$$= 6\times 10^9 \text{ years.}$$

25. ***Find the half-life of uranium, given that 3.23×10^{-7} gm of radium is found per gm of uranium in old minerals. The atomic weights of uranium and radium are 238 and 226 and half-life of radium is 1600 years. (Avogadro Number is 6023 × 10^{23}/gm-atom).***

Solution: The half-life of uranium is very much longer than that of radium. Therefore, in very old minerals, radioactive equilibrium exists, and we have

where N_U and N_R are the number of atoms of uranium and radium at any time, and λ_U and λ_R the corresponding decay constants.

Since $\lambda \propto \frac{1}{T}$ (where T is half-life), we can write

$$\frac{N_U}{T_U} = \frac{N_R}{T_R}$$

or $$T_U = \left(\frac{N_U}{N_R}\right). \qquad (i)$$

N_U is the number of uranium atoms in *l* gm of uranium while N_R is the number of radium atoms in 3.23×10^{-7} gm of radium. Thus

$$Nu = \frac{6.023 \times 10^{23}}{238}$$

and $$N_R = \frac{6.023 \times 10^{23}}{226} \times (3.23 \times 10^{-7}).$$

$$\therefore \quad \frac{N_U}{N_R} = \frac{226}{238 \times 3.23 \times 10^{-7}} = 2.94 \times 10^6.$$

Substituting for $\frac{N_U}{N_R}$ and for T_R (= 1600 years) in eq. (*i*), we get

$$T_U = (1600 \text{ years}) \times (2.94 \times 10^6)$$

$$= 4.70 \times 10^9 \text{ years.}$$

26. ***The atomic ratio between the uranium isotopes U^{238} and U^{234} in a mineral sample is found to be 1.8×10^4. The half-life of U^{234} is 2.5×10^5 years. Find the half-life of U^{238}***

[Ans. 4.5×10^9 years**].**

7

The Coherences

A wave which appears to be a pure sine wave for an infinitely large period of time or in an infinitely extended space is said to be a perfectly coherent wave. In such a wave there is a definite relationship between the phase of the wave at a given time and at a certain time later, or at a given point and at a certain distance away. No actual light source, however, emits a perfectly coherent wave. Light waves which are pure sine waves only for a limited period of time or in a limited space are partially coherent waves.

There are two different criteria of coherence; the criteria of time and the criteria of space. This gives rise to temporal coherence and spatial coherence.

Temporal Coherence

The oscillating electric field *E* of a perfectly coherent light wave would have a constant amplitude of vibration at any point, while its phase would vary linearly with time. As a function of time the field would appear as shown in Fig. It is an ideal sinusoidal function of time.

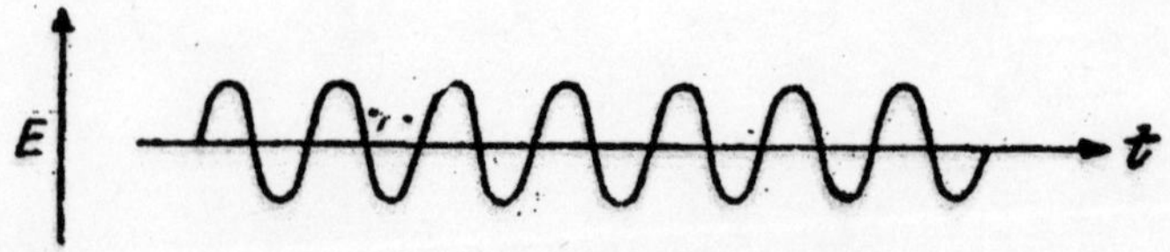

However, no light emitted by an actual source produces an ideal sinusoidal field for all values of time. This is because when an excited atom returns to the initial state, it emits light "pulse" of short duration such as of the order of 10^{-10} second for sodium atom. Thus, the field remains sinusoidal for time-intervals of the order of 10^{-10} second, after which the phase changes abruptly. Hence the field due to an actual light source will be as shown in Fig. The average time-interval for which the field remains sinusoidal (*i.e.* definite phase relationship exists) is known as "coherence time" or "temporal coherence" of the light beam, and is denoted by τ. The distance L for which the field is sinusoidal is given by.

$$L = \tau c,$$

where c is the speed of light. L is called the "coherence length" of the light beam.

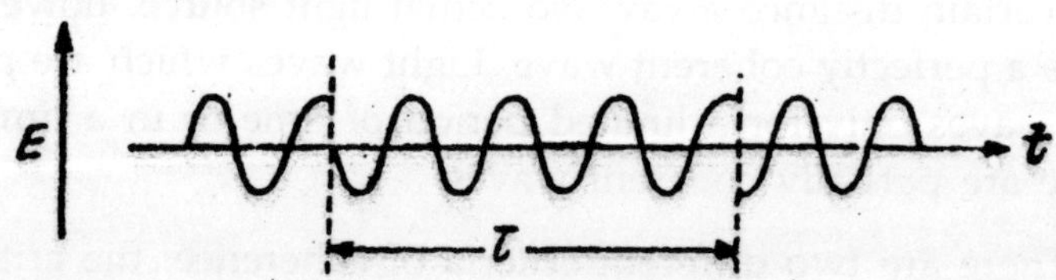

The coherence time (or the coherence length) can be measured by means of Michelson's interferometer (Fig.). A light beam from the source S falls on a semi silvered plate P at which it is partly reflected and partly transmitted. The reflected and transmitted beams, 1 and 2, are reflected back from mirrors M_1 and M_2 respectively and enter the telescope T in which interference fringes are observed. We know that the two beams can produce a stationary interference pattern only if there is a definite phase relationship between them.

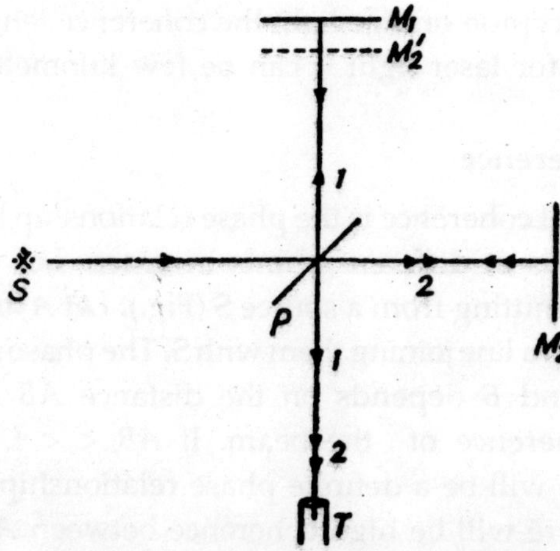

Let M_z ' be the image of M_2 formed by the plate *P*. The arrangement is then equivalent to an air-film enclosed between two reflecting surfaces M_1 and M_2 '. If *d* is the separation between M_1 and M_2', then 2*d* will be the path difference between the interfering beams. Now, if

$$2d < < L,$$

then there will be a definite phase relationship between the two beams and interference fringes will be observed. If, on the other hand,

$$2d > > L,$$

there will be no definite phase relationship and the fringes will not be observed. Therefore, starting with equal path-lengths, as the distance *d* is increased (by moving one mirror), the fringes become gradually poorer in contrast and finally disappear. The path difference at disappearance gives an estimation of coherence length.

For the sodium yellow light the coherence length is about 3 cm, so that the coherence time is

$$\tau = \frac{L}{c} \approx \frac{3\text{cm}}{3 \times 10^{10}\ \text{cm / sec}} \approx 10^{-10}\ \text{sec.}$$

For the Krypton orange light the coherence length is about 30 cm while for laser light it can be few kilometers.

Spatial Coherence

The spatial coherence is the phase relationship between the radiation fields at different points in space. Let us consider light waves emitting from a source S (Fig.). Let A and B be two points lying on a line joining them with S. The phase relationship between A and B depends on the distance AB and on the temporal coherence of the beam. If $AB << L$ (coherence length), there will be a definite phase relationship between A and B i.e. there will be high coherence between A and B. On the other hand, if $AB >> L$, there will be no coherence between A and B.

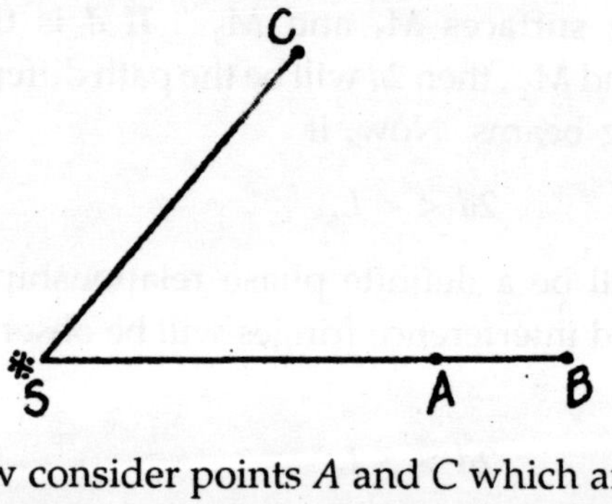

Let us now consider points A and C which are equidistant from S. If the source S is a true *point* source, then the waves shall reach A and C in exactly the same phase *i.e.* the two points will have perfect (spatial) coherence. If, however, the source S is extended, the points A and C will no longer remain in coherence. This may be emonstrated by Young's double-slit experiment illustrated in Fig. Light emitting from a narrow slit S falls on two slits S_1 and S_2 placed symmetrically with respect to S. The beams emerging from S_1 and S_2, having been derived from the same original beam, maintains a constant phase difference at all points on the screen. Hence a stationary interference pattern is observed on the screen.

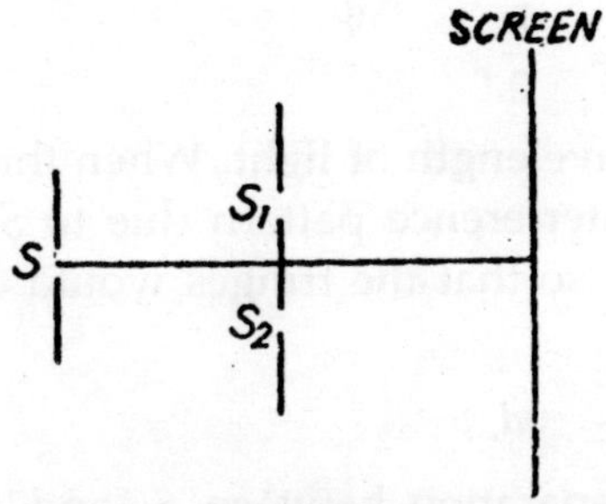

If, however, the width of the slit S is gradually increased, the pattern becomes poorer and poorer in contrast and finally disappears. This means that as the size of the source is increased, the situation of spatial coherence on the screen changes into a situation of incoherence. This happens because when S is wide, S_1 and S_2 receive waves; from different independent parts of S and hence do not remain coherent with respect to each other.

We may derive a relationship between the spatial coherence and the size of source. An extended source is made up of a largo number of point-sources. Let us first consider the case when the Young's double slit is illuminated by two independent point-sources S and S' at a distance l apart (Fig.). We shall find minimum value of l at which pattern on the screen would disappear. The waves from S which reach the point O on the screen *via* S_1 and S_2. have zero path difference. Hence there is a bright fringe at O due to S. Now, the waves from S' reaching the point O *via* S_1 and S_2 have a path difference $S'S_2$ -$S'S_1$ = KS_2. Clearly, we shall obtain a dark fringe at O due to S' when

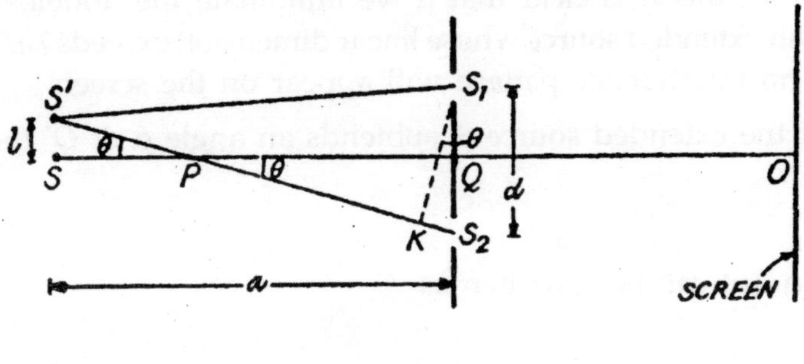

$$KS_2 = \frac{\lambda}{2}, \qquad \text{... } (i)$$

where λ is the wavelength of light. When this is the case, the maxima of the interference pattern due to S will fall on the minima due to S' so that the fringes would disappear. Now, from the figure,

$$KS_2 \simeq \theta d,$$

where d is the separation between S_1 and S_2. Let a be the distance between S and Q. Now, again from the figure

$$\theta \simeq \frac{QS_2}{PQ} \simeq \frac{SS'}{SP}.$$

$$\therefore \quad a = SP + PQ \simeq \frac{SS'}{\theta} + \frac{QS_2}{\theta} = \frac{l}{\theta} + \frac{d/2}{\theta}$$

or $$\theta \simeq \left(l + \frac{d}{2}\right)/a.$$

$$\therefore \quad KS_2 \simeq \theta d \simeq \left(l + \frac{d}{2}\right) d/a.$$

Assuming that $l >> \frac{d}{2}$; we can write

Thus in view of eq. (*i*), the interference pattern would disappear if

$$\frac{ld}{a} \simeq \frac{\lambda}{2}$$

or $$\lambda \sim \frac{\lambda a}{2d} \qquad \text{... } (ii)$$

From this it is clear that if we illuminate the double-slit with an extended source whose linear dimension exceeds $\lambda a/2d$, then no interference pattern will appear on the screen.

If the extended source S' subtends an angle α at Q' then

$$\propto \simeq \frac{l}{a}$$

Eq. (*ii*) can be rewritten as

$$d \simeq \frac{\lambda a}{2l} \simeq \frac{\lambda}{2\alpha} .$$

$\frac{\lambda}{\alpha}$ is termed as "lateral spatial coherence width". We conclude that to obtain a good interference pattern with Young's double-slit, the separation between the slits (S_1 and S_2) should be kept much less than the coherence width.

Purity of a Spectral Line : Every spectral line has a finite width which means that it corresponds to a continuous distribution of wavelengths in some narrow interval between λ and . $\lambda + \Delta\lambda$. This is obvious from the fact that for every spectral line the interference pattern in the Michelson interferometer experiment eventually disappears when the path difference between the interfering beams is gradually increased. We also know that the pattern disappears when the path difference exceeds the coherence length. Thus the concept of temporal coherence is directly related to the width (or purity) of the spectral line.

Let us adopt the criterion that when the path difference between the interfering beams becomes equal to the coherence length *L*, the rings due to two closely spaced wavelengths λ_1 and λ_2 are completely out of step at the centre (*i.e.* a bright ring of λ_1 coincides with a dark ring of λ_2). Then we can write

$$L = n\lambda_1 = (n + \frac{1}{2})\ \lambda_2.$$

Eliminating *n*, we get

$$L = \left(\frac{L}{\lambda_1} + \frac{1}{2}\right)\lambda_2$$

or $$\frac{L}{\lambda_2} - \frac{L}{\lambda_1} = \frac{1}{2}$$

or $$L\ \frac{\lambda_1 - \lambda_2}{\lambda_1\lambda_2} = \frac{1}{2}$$

or $$L = \frac{\lambda_1\lambda_2}{2(\lambda_1 - \lambda_2)}$$

If instead of two discrete wavelengths λ_1 and λ_2, the beam consists of all wavelengths lying between λ_1 and λ_2, then the pattern would disappear if

$$L = \frac{\lambda_1 \lambda_2}{\lambda_1 - \lambda_2}$$

or $$L \simeq \frac{\lambda^2}{\Delta\lambda}$$

or $$\Delta\lambda \simeq \frac{\lambda^2}{L}$$

Thus if the fringes become indistinct when the path difference exceeds L , we can conclude that the spectral line (of mean wavelength λ) has a wavelength-spread (width) given by $$\Delta \lambda \simeq \frac{\lambda^2}{L}.$$

For the cadmium red line ($\lambda = 6438$ Å) L is as large as about 30 cm. This corresponds to a wavelength-spread given by

$$\Delta\lambda \simeq \frac{\lambda^2}{L} \simeq \frac{\left(6438 \text{ Å}\right)^2}{30 \times 10^8 \text{ Å}} \simeq 0.01 \text{ Å}.$$

Further, since $v = \dfrac{c}{\lambda}$, the frequency-spread Dv of a line would be

$$\Delta v = \frac{c}{\lambda^2} \Delta\lambda \simeq \frac{c}{L}$$

We know that $\tau \simeq \dfrac{L}{c}$ where τ is coherence time.

$$\therefore \qquad \Delta v \sim \frac{1}{\tau}.$$

Thus the frequency-spread of a spectral line is of the order of the inverse of the coherence time. It means that a perfectly sharp monochromatic line ($\Delta v - 0$) would correspond to an infinite interval of time ($\tau = \infty$).

8

The Laser

The word 'LASER' is an acronym for Light Amplification by Stimulated Emission of Radiation. It is a device to produce a strong, monochromatic, collimated and highly coherent beam of light; and depends on the phenomenon of "stimulated emission", first predicted by Einstein in 1916. Einstein considered the equilibrium between matter and electromagnetic radiation in a black-body chamber at a constant temperature in which exchange of energy takes place due to absorption and spontaneous emission of radiation by the atoms. He found that the usual absorption and emission processes alone are not sufficient to explain the equilibrium. He then predicted that there must be a third process also, now called "stimulated emission". This prediction was paid little attention until 1954, when Townes and Gordon developed a microwave amplifier (MASER) using ammonia, NH_3. In 1958, Schawlow and Townes showed that the maser principle could be extended into the visible region and in 1960, Maiman built the first laser using ruby as the active medium. Since then laser has opened up completely new fields of development in optics.

Absorption of Radiation

An atom has a number of possible quantised energy states characterised by integral numbers. If it is initially in a lower state 1, it can rise to a higher state 2 by absorbing a quantum of radiation (photon) of frequency v, given by

$$v = \frac{E_2 - E_1}{h},$$

where E_1 and E_2 are the energies of the atom in the states 1 and 2 respectively (Fig.). This is absorption of radiation. This is a stimulated (or induced) process, the absorbed photon being the stimulating photon. (The absorption is necessarily stimulated).

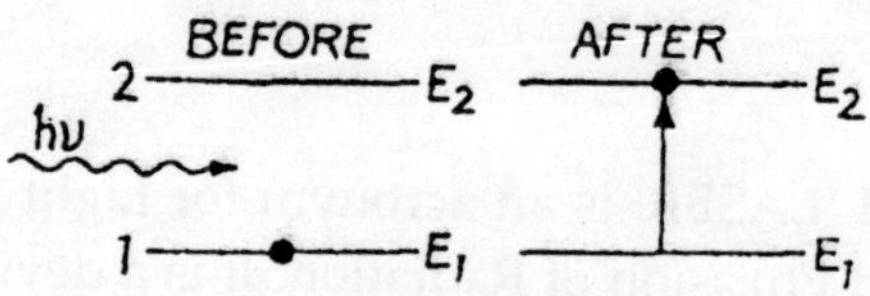

(ABSORPTION)

The probable rate of occurrence of this absorption transition 1→2 depends on the properties of states 1 and 2 and is proportional to the energy density $u(v)$ of the radiation of frequency v incident on the atom. Thus

$$P_{12} = B_{12}\ u(v). \qquad \text{... }(i)$$

The proportionality constant B_{12} is known as 'Einstein's coefficient of absorption of radiation'.

Spontaneous Emission

Let us now consider an atom initially in the higher (excited) state 2 (Fig.). Observations show that its life-time in higher state is usually very small C ≈ 10^{-8} second) and it, of its own accord, jumps to the lower energy state 1, emitting a photon of frequency v. This is 'spontaneous' emission of radiation. If there is an assembly of atoms, the radiation emitted spontaneously by each atom has a random direction and a random phase, and is therefore incoherent from one atom to another.

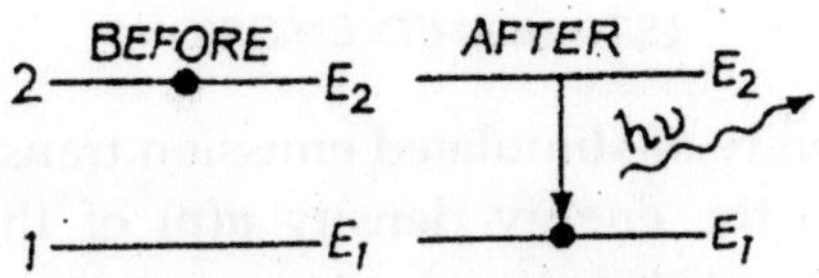

(SPONTANEOUS EMISSION)

The probability of spontaneous emission 2 → 1 is determined only by the properties of states 2 and 1. Einstein denoted this probability per unit time by

$$A_{21}$$

which is known as 'Einstein's coefficient of spontaneous emission of radiation'.

We note that the probability of absorption transitions depends upon the energy density «(v) of the incident radiation, whereas the spontaneous emissions are independent of it. Hence, for equilibrium, emission transitions depending upon $u(v)$ must also exist. These are 'stimulated' emission transitions.

Stimulated (or Induced) Emission : According to Einstein, an atom in an excited energy state may, under the influence of the electromagnetic field of a photon of frequency v incident upon it, jump to a lower energy state, emitting an additional photon of same frequency v (Fig.). Thus now two photons, one original and the other emitted, move on. This is 'stimulated' emission of radiation (or negative absorption of radiation). The direction of propagation, energy, phase and state of polarisation of the emitted photon is exactly the same as that of the incident stimulating photon. In other words, the stimulated radiation is completely coherent with the stimulating radiation. As a result of this process, radiation passing through an assembly of atoms is amplified.

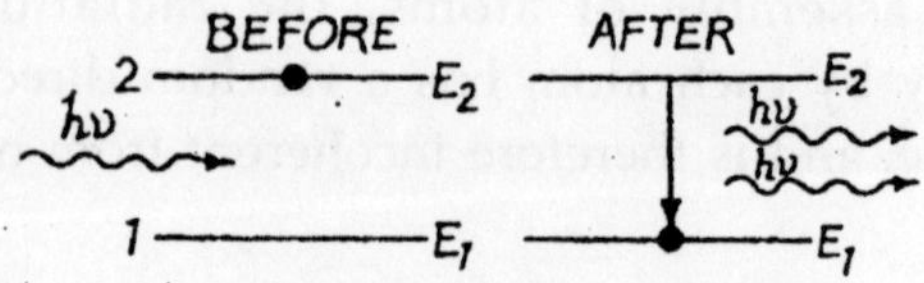

(STIMULATED EMISSION)

The probability of stimulated emission transition 2 → 1 is proportional to the energy density $u(v)$ of the stimulating radiation and is written as

$$B_{21}\ u(v),$$

where B_{21} is the 'Einstein's coefficient of stimulated emission of radiation'.

The total probability for an atom in state 2 to drop to the lower state 1 is therefore

$$P_{21} = A_{21} + B_{21}\ u(v). \qquad \text{... } (ii)$$

Relation between Spontaneous and Stimulated Emission Probabilities : Let us consider an assembly of atoms in thermal equilibrium at temperature T with radiation of frequency v and energy density $u(v)$. Let N_1 and N_2 be the number of atoms in states 1 and 2 respectively at any instant. The number of atoms in state 1 that absorb a photon and rise to state 2 per unit time is

$$N_1P_{12} = N_1B_{12}\ u(v). \qquad \text{[by eq. } (i)]$$

Conversely, the number of atoms in state 2 that drop to 1, either spontaneously or under stimulation, emitting a photon per unit time is

$$N_2P_{21} = N_2\ [A_{21} + B_{21}\ u(v)] \qquad \text{[by eq. } (ii)]$$

For equilibrium, the absorption and emission must occur equally.

Thus

$$N_1P_{12} = N_2P_{21}$$

$$\text{or} \quad N_1 B_{12}\, u(v) = N_2\, [A_{21} + B_{21}\, u(v)$$

$$\text{or} \quad u(v) = \frac{N_2 A_{21}}{N_1 B_{12} - N_2 B_{21}}$$

$$\text{or} \quad u\,(v) = \frac{A_{21}}{B_{21}} \frac{1}{\frac{N^1}{N_2}\left(\frac{B_{12}}{B_{21}}\right) - 1}$$

Einstein proved thermodynamically that the probability of (stimulated) absorption is equal to the probability of stimulated emission, *i.e.*

$$B_{12} = B_{21}$$

Then, we have

$$u\,(v) = \frac{A_{21}}{B_{21}} \frac{1}{\frac{N_1}{N_2} - 1}$$

The equilibrium distribution of atoms among different energy states is given by Boltzmann's law according to which

$$\frac{N_2}{N_1} = \frac{e^{-E_2/kT}}{e^{-E_1/kT}}$$

$$\text{or} \quad \frac{N_1}{N_2} = e^{-(E_2 - E_1)/kT} = e^{hv/\kappa T}$$

$$\text{Consequently,} u(v) = \frac{A_{21}}{B_{21}} \frac{1}{e^{hv/kT}_{-1}}$$

This is a formula for the energy density of photon of frequency v in equilibrium with atoms in energy states 1 and 2, at temperature *T*. Comparing it with the Planck radiation formula

$$u\,(v) = \frac{8\pi h v^3}{c^3} \frac{1}{e^{hv.kT}_{-1}},$$

we get

$$\boxed{\frac{A_{21}}{B_{21}} = \frac{8\pi h v^3}{c^3}}$$

This is the formula for the ratio between the spontaneous emission, and induced emission coefficients. This ratio is proportional to v^3. It means that the probability of spontaneous emission increases rapidly with the energy difference between two states.

Features of Stimulated Radiation

Let us consider an ensemble of atoms irradiated with light of frequency v which coincides with one of the characteristic frequencies of the atoms. At room temperature, or higher, a certain number of atoms will be in an excited state. Now, two processes may occur : (*i*) an absorption transition of atoms from a lower energy level 1 to a higher energy level 2, such that $v = (E_2 - E_1,)/h$; (*ii*) stimulated emission transition from higher energy level 2 to lower energy level 1.

In the first process, a photon from the incident beam is absorbed by an atom, thus leading the atom to an excited state. In the second process, an incident photon forces the excited atom to emit another photon of the same frequency in the same direction and in the same phase. The two photons go off together as coherent radiation.

Requirements for Laser Action : Under ordinary conditions of thermal equilibrium the number of atoms in higher energy state 2 is considerably smaller than the number in lower energy state 1 $(N_2 < N_1)$, so that there is very little stimulated emission compared with absorption. An incident photon is more likely to be absorbed than to cause emission.

If, however, by some means a larger number of atoms are made available in the higher energy state, stimulated emission is promoted. The situation in which the number of atoms in the higher energy state exceeds that in the lower state $(N_2 > N_1)$ is known as "population inversion". In this situation the system of atoms would lase.

Stimulated emissions are further encouraged by increasing the radiation density $u(v)$ of the stimulating radiation. This is achieved by enclosing the emitted radiation in a "cavity" between two parallel reflectors. The radiation repeatedly travels back and forth, and the photons passing through the atoms go on multiplying by repeated stimulated emission (Fig.). Hence a strong coherent beam of light emerges from the system. (Fig.)

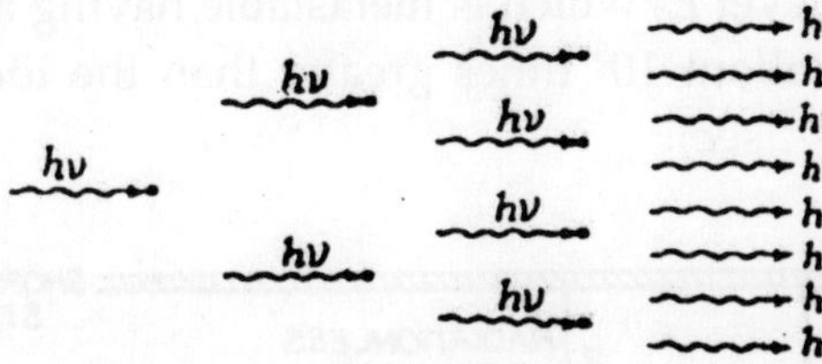

Pumping : The process of achieving population inversion is known as "pumping" of atoms. There are various types of pumping process, but the most natural is the 'optical pumping' which is utilised in Ruby laser.

The Ruby Laser : This is the first laser developed in 1960, and is a solid-state laser. It consists of a pink ruby cylindrical rod whose ends are optically flat and parallel (Fig.). One end is fully silvered and the other is only partially silvered. Upon the rod is wound a coiled flash lamp filled with xenon gas.

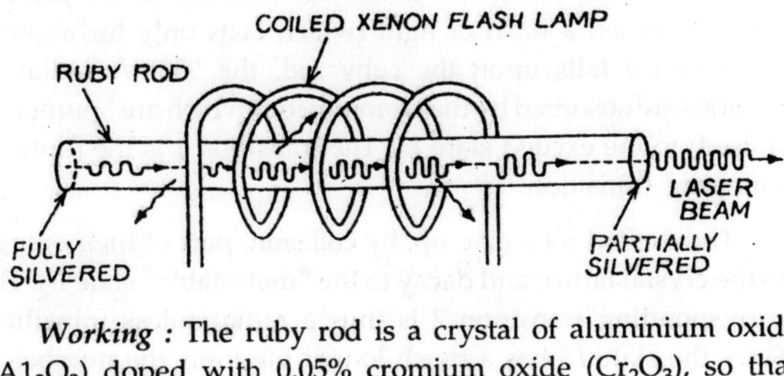

Working : The ruby rod is a crystal of aluminium oxide ($A1_2O_3$) doped with 0.05% cromium oxide (Cr_2O_3), so that some of the $A1^{+++}$ ions are replaced by Cr^{+++} ions. These "impurity" chromium ions give pink colour to the ruby and give rise to the laser action.

In Fig. is shown a simplified version of the energy-level diagram of chromium ion. It consists of an upper short-lived energy level (rather energy band) E_3 above its ground-state energy level E_1, the energy difference E_3—E_1 corresponding to a wavelength of about 5500 A. There is an intermediate excited-state level E_2 which is metastable having a life-time of 3×10^{-3} sec (about 10^5 times greater than the life-time of E_3 which is $\approx 10^{-8}$ sec).

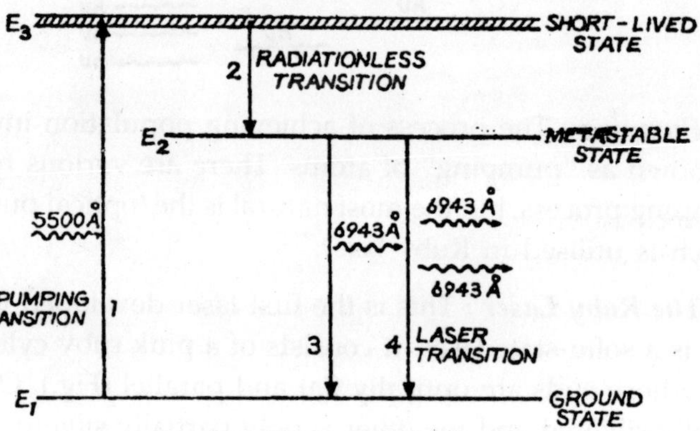

A THREE-LEVEL LASER

Normally, most of the chromium ions are in the ground state E_1. When a flash of light (which lasts only for about a millisecond) falls upon the ruby rod, the 5500-Å radiation photons are absorbed by the cromium ions which are "pumped" (raised) to the excited state E_3. The transition 1 is the (optical) pumping transition.

The excited ions give up, by collision, part of their energy to the crystal lattice and decay to the "metastable" state E_2. The corresponding transition 2 is thus a radiationless transition. Since the state E_2 has a much longer life-time, the number of ions in this state goes on increasing while, due to pumping, the number in the ground state E_1 goes on decreasing. Thus population inversion is established between the metastable (excited) state E_2 and the ground state E_1

When an (excited) ion passes spontaneously from the metastable state to the ground state (transition 3), it emits a photon of wavelength 6943 Å. This photon travels through the ruby rod and, if it is moving parallel to the axis of the crystal, is reflected back and forth by the silvered ends until it stimulates an excited ion and causes it to emit a fresh photon in phase with the stimulating photon. This "stimulated" transition 4 is the laser transition. (The photons emitted spontaneously which do not move axially escape through the sides of the crystal).

The process is repeated again and again because the photons repeatedly move along the crystal being reflected from its ends. The photons thus multiply. When the photon-beam becomes sufficiently intense, part of it emerges through the partially-silvered end of the crystal.

There is a drawback in the three-level laser such as ruby. The laser requires high pumping power because the laser transition terminates at the ground state and more than one-half of the ground-state atoms must be pumped up to the higher state to achieve population inversion. Moreover, ions which happen to be in their ground state absorbs the 6943-Å photons from the beam as it builds up.

The ruby laser is a "pulsed" laser. The active medium (Cr^{+++} ions) is excited in pulses, and it emits laser light in pulses. While the Xenon pulse is of several millisecond duration ; the laser pulse is much shorter, less than a millisecond duration. It means enhanced instantaneous power.

Helium-Neon Laser : It is a four-level laser in which the population inversion is achieved by electric discharge. A mixture of about 7 parts of helium and 1 part of neon is contained in a glass tube at a pressure of about 1 mm of mercury. (Fig.).

At both ends of the tube are fitted optically plane and parallel mirrors, one of them being only partially silvered. The spacing of the mirrors is equal to an integral number of half-wavelengths of the laser light. An electric discharge is produced in the gas-mixture by electrodes connected to a high-frequency electric source.

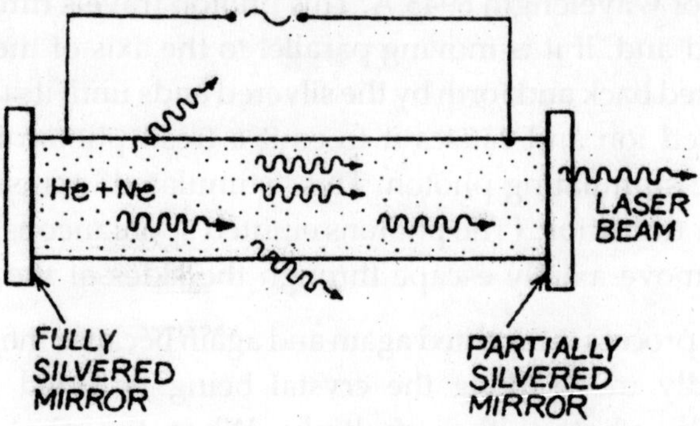

The electrons from the discharge collide with and "pump" (excite) the He and Ne atoms to metastable states 20.61 eV and 20.66 eV respectively above their ground states (Fig.). Some of the excited He atoms transfer their energy to ground-state Ne atoms by collisions, with the 0.05 eV of additional energy being provided by the kinetic energy of atoms. Thus He atoms help in achieving a population inversion in the Ne atoms.

When an excited Ne atom passes spontaneously from the metastable state at 20.66 eV to state at 18.70 eV, it emits a 6328-019Å photon. This photon travels through the gas-mixture, and if it is moving parallel to the axis of the tube, is reflected back and forth by the mirror-ends until it stimulates an excited Ne atom and causes it to emit a fresh 6328-Å photon in phase with the stimulating photon. This stimulated transition from 20.66-eV level to 18.70-eV level is the laser transition. This process is continued and a beam of coherent radiation builds up in the tube. When this beam becomes sufficiently intense, a portion of it escapes through the partially-silvered end.

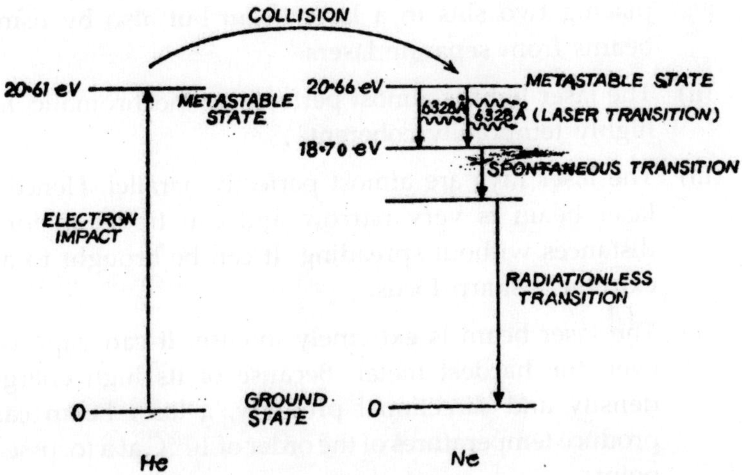

A FOUR-LEVEL LASER

From the 18.70-eV level the Ne atom passes down spontaneously to a lower metastable state emitting incoherent light, and finally to the ground state through collision with the tube walls. The final transition is thus radiationless.

Obviously, the Ne, atom in its ground state cannot absorb the 6328-A photons from the laser beam, as happens in the three-level ruby laser. Also, because the electron impacts that exite the He and Ne atoms occur all the time, unlike the pulsed excitation from the xenon flash lamp in the ruby laser, the He-Ne laser operates continuously.

Further, since the laser transition does *not* terminate at the ground state, the power needed for excitation is less than that in a three-level laser.

Properties of a Laser Beam : The laser beam has certain characteristic properties which are not present in beams derived from other light sources :

(i) The laser beam is completely spatially coherent, with the waves all exactly in phase with one another. An interference pattern can be obtained not merely by placing two slits in a laser beam but also by using beams from separate lasers.

(ii) The laser light is almost perfectly monochromatic, *i.e.* highly temporally coherent. ,

(iii) The laser rays are almost perfectly parallel. Hence a laser beam is very narrow and can travel to long distances without spreading. It can be brought to an extremely sharp focus.

(iv) The laser beam is extremely intense. It can vaporise even the hardest metal. Because of its high energy density and directional property, a laser beam can produce temperatures of the order of 10^4°C at a focussed point.

Applications of Lasers : The laser beam being narrow, intense, parallel, monochromatic and highly coherent is finding increasing applications in various fields :

(i) In the technical and industrial field, the laser beam is used for cutting fabric for clothing on one hand and steel sheets on the other. It can drill extremely fine holes in paper clips, single human hair and hard materials including teeth and diamond. Extremely thin wires used in cables are drawn through the diamond hole. Metallic rods can be melted and joined by means of a laser beam (laser welding). The surfaces of engine crankshafts and the cylinder walls are hardened through heat-treatment by laser. The laser beam is used to vaporise unwanted material during the manufacture of electronic circuits on semiconductor chips.

(ii) In the medical field, the laser beam is used in delicate surgery like cornea grafting. Using laser beam, the surgical operation is completed in a much shorter time. It is also used in the treatment of kidney stone, cancer, tumour and in cutting and sealing the small blood vessels in brain operation.

(iii) During war-time, lasers are used to detect and destroy enemy missiles. Now, laser-rifles, laser-pistols and laser-bombs are also being made which can be aimed at the enemy in the night. In space, laser has been used to control rockets and satellites and in directional radio-communication like fiberoptic telephony.

(iv) Laser is very useful in science and research. It has been used to perform Michelson-Morley experiment which is the building stone of the Einstein's theory of relativity. It can be used to determine the temperature of plasma and the density of electron. Laser-torch is used to see objects at long distances.

(v) Laser is used in holography and nonlinear optics.

(vi) Since laser rays are very much parallel, so they are used for communications and measuring long distances. The distance between earth and moon has been measured by laser rays to an accuracy of 15 cm.

(vii) Laser rays have proved to be useful in detecting nuclear explosions and earthquakes, in vaporising solid fuel of rockets, in the study of the surface of distant planets and satellites.

(viii) Laser beams have also been used in the "inertial confinement" of plasma.

(iii) During war-time, lasers are used to detect and destroy enemy missiles. Now, laser-rifles, laser-pistols and laser-bombs are also being made which can be aimed at the enemy in the night. In space, laser has been used to control rockets and satellites and in-direct-bond radio-communication like fibre-optic telephony.

(iv) Laser is very useful in science and research. It has been used to perform Michelson-Morley experiment which is the building stone of the Einstein's theory of relativity. It can be used to determine the temperature of plasma and the density of electron. Laser-torch is used to see objects at long distances.

(v) Laser is used in holography and nonlinear optics.

(vi) Since laser rays are very much parallel, so they are used for communications and measuring long distances. The distance between earth and moon has been measured by laser rays to an accuracy of 15 cm.

(vii) Laser rays have proved to be useful in detecting nuclear explosions and earthquakes, in vaporising solid fuel of rockets, in the study of the surface of distant planets and satellites.

(viii) Laser beams have also been used in the inertial confinement of plasma.

9

Periodic Table of Elements

Quantum Numbers Specifying the State of an Electron in an Atom: The study of the spectra of atoms, and of the Zeeman effect, has suggested that an electron in the atom can exist only in one of the several discrete energy states, called 'quantum states'. Each state is quantised with regard to size, shape and orientation of the electron-orbit and is described by a set of four quantum numbers. These quantum numbers are as follows :

(i) n :— It is the 'total' or 'principal' quantum number which takes integral values from 1 to infinity :

$$n = 1, 2, 3, \ldots\ldots\ldots, \infty.$$

It determines the general size of the electron-orbit and the energy of the electron.

(ii) l : — It is the 'orbital' or 'azimuthal' quantum number which takes integral values from 0 to $n - 1$:

$$l = 0, 1, 2 \ldots\ldots\ldots, n\text{-}1.$$

It determines the shape of the electron-orbit and the orbital angular momentum of the electron.

(iii) j :— It is the 'inner' quantum number which takes two values given by

$$j = l \pm s = l \pm \frac{1}{2}$$

where s is the (intrinsic) spin quantum number and takes the value $\frac{1}{2}$ for all electrons. The quantum number j determines the total (orbital + spin) angular momentum of the electron.

(iv) mj :— It is the 'magnetic' quantum number corresponding to the inner quantum number j, and takes the following values :

$$mj = j, j-1, \ j-2, \dots \dots 1, 0, -1, -2, \ -j \ ;$$

that is, a total of $(2j+1)$ values. It determines the orientation of electron-orbit in an external magnetic field.

Thus the four quantum numbers n, l, j and m_j specify completely the (quantum) state of an electron in an atom, and hence determine its energy and angular momentum. When an electron jumps from one state to another, some or all of these quantum numbers, and hence the energy and angular momentum of the electron, change. (The charge, rest mass and spin do not change as these are intrinsic properties of electron).

Shell Structure of Atom

The permissible orbits (*i.e.* quantum states) of the atom can be grouped together into 'shells'. Electrons that have the same total quantum number n are roughly, on the average, at the same distance from the nucleus and have some energies. All these electrons are said to occupy the same atomic "shell". These shells are denoted by capital letters $K, L, M, N, \dots$ according as $n = 1, 2, 3, 4, \dots\dots$ The K-shell (for which $n - 1$) is nearest the nucleus.

The energy of an electron in a particular shell also depends upon its orbital quantum number l (except for hydrogen-like atoms), though this dependence is not so great as that upon n. Therefore, each shell is considered as composed of sub-shells. The electrons in a shell having the same value of l are said to occupy the same "sub-shell." All the electrons in a sub-shell have almost identical energies. The sub-shells are denoted by small letters s, p, d, f ...... according as $l = 0, 1, 2, 3,$ .... Thus the various shells have the following sub-shells :

[Remember that l can take values from 0 to (n — 1)]

Shell	*n*	*l*	*no. of sub-shells*	*names of sub-shells*
K	1	0	1	1s
L	2	0,1	2	2*s*, 2*p*
M	3	0, 1, 2	3	3*s*, 3*p*, 3*d*,
...	...	...	...	...
...	...	...	...	...

Thus the *K*-shell has one sub-shell called is the *L*-shell has two sub-shells called 2*s* and 2*p*, and so on. In general, *a shell characterised by a total quantum number n has n sub-shells.*

Pauli's Exclusion Principle and Distribution of Electrons Among the Shells : The electrons in an atom are distributed in a definite way among the various shells and sub-shells. This distribution is governed by a principle given by Pauli in 1925. This is known as 'Pauli's exclusion principle'.

It states that *no two electrons in an atom can exist in the same quantum state.* This means that no two electrons can have the same set of the four quantum numbers n, l, j and m_j. On this basis we can deduce the number of *permitted* electrons in various sub-shells, and hence in the various shells.

In a given sub-shell all the electrons have the same values of n and l. Hence they must differ either in j or in mj. Now for a given l, there are two different values of j, namely, $l + \frac{1}{2}$ and $l - \frac{1}{2}$. Let us call them j' and j'' respectively. Further, for a given j, we have $[2j + 1]$ different values of mj. Thus for j' we have

$$[2j' + 1] = [2(l + \frac{1}{2}) + 1] = (2l + 2) \text{ values of } mj ;$$

and for j'' we have

$$[2j'' + 1] = 2(l - \frac{1}{2}) + 1] = 2l \text{ values of } mj.$$

This means that in a given sub-shell we can have $(2l + 2) + 2l$ electrons which will differ either in j or in m_j. Hence the number of maximum electrons permissible in a sub-shell

$$= (2l + 2) + 2l = 4l + 2 = 2(2l + 1).$$

On this basis the various sub-shells can have the following maximum electrons :

Sub-shell	*l*	*maximum electrons* 2(2*l*+1)
s	0	2
p	1	6
d	2	10
f	3	14
...	...	...

Thus an *s* sub-shell can have a maximum of 2 electrons, a *p* sub-shell can have a maximum of 6 electrons, and so on.

Now, a given shell has n sub-shells, with l having integral values from 0 to $(n - 1)$. Hence the maximum number of electrons in a shell

$$= \sum_{l=0}^{n-1} 2(2l+1)$$

$$= 2[1 + 3 + 5 + 7 + \dots + 2(n-1) + 1]$$

$$= 2[1 + 3 + 5 + 7 + \dots \dots(2n-1)\ ]$$

$$= \left[\frac{n}{2}\left\{1+(2n-1)\right\}\right]$$

$$= 2n^2.$$

Thus the *K*-shell ($n = 1$) can have a maximum of 2 electrons, the *L*-shell ($n = 2$) a maximum of 8 electrons, the *M*-shell ($n = 3$) a maximum of 18 electrons, and so on. The final distribution can be tabulated as below :

Shell	*n*	*l*	*Sub-shells*	*Maximum Electrons in each sub-shell 2(2l + 1)*	*Total $2n^2$*
K	1	0	1*s*	2	2
L	2	0, 1	2*s*, 2*p*	2, 6	8
M	3	0, 1, 2	3*s*, 3*p*, 3*d*	2, 6, 10	18
N	4	0, 1, 2, 3	4*s*, 4*p*, 4*d*, 4*f*	2, 6, 10, 14	32
...	...	...	...	...	...

A shell or sub-shell having its full quota of electrons is said to be 'closed' or 'filled'.

Magnetic Moment of a Closed Shell: Every electron in a closed sub-shell is paired with another electron having the same value of m_j but of opposite sign. For example, in a 2*p* sub-shell ($n = 2$, $l = 1$), the 6 electrons have the following values of j and m_j:

j	$\frac{1}{2}$	$\frac{3}{2}$
m_j	$\frac{1}{2}, -\frac{1}{2}$	$\frac{3}{2}, \frac{1}{2}, -\frac{1}{2}, -\frac{3}{2}$

Thus for all the electrons, we have

$$\Sigma m_j = 0,$$

that is, for the (closed) sub-shell the total (orbital plus spin) angular momentum in a specified direction is zero. As a result, a closed shell, which is made up of closed sub-shells, does not contribute toward the total angular momentum, and hence toward the magnetic moment, of the atom.

Pauli's Exclusion Principle has been stated and explained earlier. If applies to those particles which obey Fermi-Dirac statistics and are known as "Fermions". They can be described only by "anti-symmetric" total wave-functions. Such particles have half-integral spins, such as electron, proton, neutron, etc.

Building-up of Electronic Configurations in Atoms : We can deduce the electronic configuration of atoms by filling up the sub-shells and shells with electrons turn by turn, starting with the most strongly bound *K*-shell. If the nuclear charge of an atom is *Ze,* then Z is called its 'atomic number' and Z electrons are to be filled in the shells of the atom. The filling up is done in an order such that all the electrons of the atom are in the lowest energy states available to them and that any sub-shell is not filled beyond its capacity of $2(2l + 1)$ electrons. In this condition the atom is most stable and is said to be in the 'ground' state.

Let us now consider the atoms of a few elements :

Hydrogen (Z = 1) :— It has only one electron. This electron will go into the lowest-energy sub-shell 1s. The electronic configuration of hydrogen in the ground state is therefore

$$1s^1.$$

Helium [Z = 2) : — It has two electrons, both of which can go into the 1s sub-shell. The energy of the atom will then be a minimum. The configuration of the helium atom is thus

$\boxed{1s^2}$

The *K*-shell is closed in the helium atom.

Lithium (Z = 3):—It has three electrons. Of these, two go into the 1s sub-shell, when the *K*-shell is closed. The remaining third electron will then go to the next available lowest-energy sub-shell 2s of the *L*-shell. The configuration of the Li atom is thus

$$\boxed{1s^2}\,2s^1.$$

Neon (Z = 10) :— It has 10 electrons. In order that the atom has a minimum energy, the first two electrons will go into the 1s sub shell, the next two in the 2s sub-shell, and the remaining six in the *2p* sub-shell. Thus 2 electrons occupy the k-shell and 8 electrons occupy the *L*-shell. The configuration is

$$\boxed{1s^2}\;\boxed{2s^2\,2p^6}$$

The *L*-shell is also closed in the Ne atom.

Sodium (Z = 11) — It has 11 electrons. Of these, the first two go into the 1*s* sub-shell thus closing she *K*-shell, the next two go into the 2*s* sub-shell and then six go into the 2*p* sub-shell thus closing the *L*-shell also. The remaining eleventh electron goes to the next available sub-shell 3*s* of the *M*-shell. Under this condition the atom is in the lowest energy (ground) state and its electronic configuration is

$$\boxed{1s^2}\;\boxed{2s^2\,2p^6}\;3s_1.$$

For atoms with Z > 18 the energy-ordering of the sub-shells becomes somewhat irregular and is deduced from detailed study of the atomic spectra. However, once the electronic configuration has been established, many of the chemical and optical properties of the element can be understood.

The Pauli's exclusion principle and the shell structure of atoms have been discussed in the last question.

The Periodic Table

The periodic table is an arrangement of elements listed in the order of atomic number; It has the characteristic that the elements with similar chemical and physical properties recur at regular intervals. The table was first constructed by Mendeleev in 1871 and perfected by Moseley and others from a study of *X*-ray spectra. A modern version of the periodic table of elements is shown in Fig.

The table consists of seven horizontal rows which are called "periods", having 2, 8, 8, 18, 18, 32 and 17 elements respectively. The last period is rather incomplete. Across each period there is a steady transition from an active metal to an inert gas. As a consequence, elements with similar properties occur in vertical columns which are called "groups". For instance, group I consists of hydrogen plus the alkali metals ail of which are highly active while, the (last) group VIII consists of noble gases all of which are inert.

In each period after the third, a series of "transition elements" appears between the group II and group III elements. The elements of each series show chemical resemblance to one another, but are different from the elements of the preceding periods. The transition elements of period six include 14 "rare earth" elements called 'lanthanides' which are virtually indistinguishable in their properties. A similar group of closely related metals, called 'actinides' occurs in period seven.

The structure of the periodic table can be explained on the basis of the shell structure of atoms and the distribution of electrons among them. The distribution is done under Pauli's principle and is in an order such that all the electrons of the atom are in the lowest energy states available to them. The atom is then maximum stable.

PERIOD	GROUP I	II	TRANSITION ELEMENTS										GROUP III	IV	V	VI	VII	VIII
1	1 H																	2 He
2	3 Li	4 Be											5 B	6 C	7 N	8 O	9 F	10 Ne
3	11 Na	12 Mg											13 Al	14 Si	15 P	16 S	17 Cl	18 A
4	19 K	20 Ca	21 Sc	22 Ti	23 V	24 Cr	25 Mn	26 Fe	27 Co	28 Ni	29 Cu	30 Zn	31 Ga	32 Ge	33 As	34 Se	35 Br	36 Kr
5	37 Rb	38 Sr	39 Y	40 Zr	41 Nb	42 Mo	43 Tc	44 Ru	45 Rh	46 Pd	47 Ag	48 Cd	49 In	50 Sn	51 Sb	52 Te	53 I	54 Xe
6	55 Cs	56 Ba	57 La*	72 Hf	73 Ta	74 W	75 Re	76 Os	77 Ir	78 Pt	79 Au	80 Hg	81 Tl	82 Pb	83 Bi	84 Po	85 At	86 Rn
7	87 Fr	88 Ra	89 Ac†															

LANTHANIDES* (RARE EARTHS)	58 Ce	59 Pr	60 Nd	61 Pm	62 Sm	63 Eu	64 Gd	65 Tb	66 Dy	67 Ho	68 Er	69 Tm	70 Yb	71 Lu
ACTINIDES†	90 Th	91 Pa	92 U	93 Np	94 Pu	95 Am	96 Cm	97 Bk	98 Cf	99 Es	100 Fm	101 Md	102 No	103 Lw

The numerical position of the atom in the table (which is the atomic number *Z*) gives the number of positive charges in the nucleus of the atom, and is equal to the number of electrons around the nucleus. It is seen that the elements in the same period of the table have their electrons distributed among the same shells, and that each period starts with the filling-up of a new shell.

First Period: Let us start with hydrogen (Z = 1). It has only one electron which occupies the lowest-energy sub-shell 1*s*. The next element Helium (Z = 2) has two electrons both of which can go into the 1*s* sub-shell. Its configuration is therefore $\boxed{1s^2}$, and with this the *K*-shell is closed. Since the *K*-shell can hold only two electrons, there are only two elements in the first period.

Second Period: With Lithium (Z = 3) starts the second period. Its two electrons occupy the *K*-shell, and the third goes to the next available sub-shell 2*s* of the *L*-shell. Its configuration is $\boxed{1s^2}\ 2s^1$. The next element is Beryllium (Z = 4) whose configuration is $\boxed{1s^2}\ 2s^2$. The sub-shell 2*s* is now closed. In the next element Boron (Z = 5) the fifth electron goes to the next available sub-shell 2*p* so that the electron configuration is $\boxed{1s^2}\ 2S^2\ 2p^1$. The. process continues until at Neon (Z = 10) the configuration becomes $\boxed{1s^2}\ \boxed{2s^2\ 2p^6}\ 3s^1$, and the *L*-shell is closed. With this the second period is also closed.

Third Period: With Sodium (Z = 11) starts the third period. Out of its eleven electrons, ten electrons fill the *K*- and *L*-shell, and the eleventh goes to the next available sub-shell 3*s* of the *M*-shell. Its configuration is $\boxed{1s^2}\ \boxed{2s^2\ 2p^6}\ 3s^2$. This process continues until at Argon (Z = 18), the configuration becomes $\boxed{1s^2}\ \boxed{2s^2\ 2p^6}\ 3s^2\ 3p^6$, and the 3*p* sub-shell of the *M*-shell is closed (The 3*d* sub-shell of the *M-shell* is vacant). With this the third period is also closed.

With Potassium (Z = 19) and Ca (Z = 20) starts the fourth period and the filling-up of the *N*-shell. The additional electrons of these atoms go to the 4*s* sub-shell of the *N*-shell and not to the already-vacant 3*d* sub-shell of the *M*-shell. This is because 4s is lower in energy than the *3d*. The 3*d* sub-shell is filled successively in the next ten elements from Sc (Z = 21) to Zn (Z = 30). These ten elements form the first series and are called 'transition elements'. After Zn, the filling up of the 4*p* sub-shell starts. The rest of the table is constructed in a similar way. The rare-earth elements from Z = 58 to Z = 71, and then again from Z = 90 to Z = 103, are formed again in a special way.

The periodic table having built up, we can interpret some of its prominent features in terms of the electronic configuration of atoms :

(i) *Similarity of Group Elements :* Elements in the same vertical column (group) show similar chemical properties and give similar spectra. This is because they have similar electron configurations. Take, for example, the alkali metals ; Li, Na, K, etc., in the first group which are very much chemically and optically alike. All of them have closed shells or sub-shells with a single outer electron in an *s* sub-shell. Hence they show electropositive behaviour, that is, they readily give up their outermost electron (thus becoming a positive ion) and acquire a very stable configuration like that of an inert gas. This is why these alkali elements are chemically very much active.

Li (Z = 3) : $\boxed{1s^2}\ 2s^1$

Na(Z= 11) : $\boxed{1s^2}\ \boxed{2s^2\,2p^6}\ 3s^1$

K (Z = 19) : $\boxed{1s^2}\ \boxed{2s^2\,2p^6}\ \boxed{3s^2\,3p^6}\ 4s^1$

Similarly, on the other hand, the elements of the VII group, F, Cl, Br, etc. show electro-negative behaviour. This is because in all of them there is an outermost *p* sub-shell having 5 electrons. Hence these atoms leadily capture one electron from outside (thus becoming a negative ion) so that their *p* sub-shell becomes closed and acquire a very stable configuration like that of an inert gas.

F(Z = 9) : $\boxed{1s^2}\ 2s^2\ 2p^5$

Cl (Z = 17) : $\boxed{1s^2}\ \boxed{2s^2\,2p^6}\ 3s^2\ 3p^5$

From this we conclude that the electronic configuration of an atom plays the leading role in determining its properties.

(ii) *Ionisation of Atoms :* If an atom is ionised then its spectrum becomes similar to that of the atom directly preceding it in the periodic table. For example, the spectrum of Be^+ ion is similar to that of Li atom. This is because on losing of Be one electron, the configuration of Be^+ becomes the same as that of Li.

(iii) *Noble Gases :* All the elements in the last group are noble gases He, Ne, A,.........which are particularly stable and chemically inert. This is because in all of them either a shell or a *p* sub shell is closed, and after each of them a new period starts and a new shell begins to be filled up.

He (Z = 2) : $\boxed{1s^2}$

Ne (Z = 10) : $\boxed{1s^2}$ $\boxed{2s^2\,2p^6}$

Ne (Z = 18) : $\boxed{1s^2}$ $\boxed{2s^2\,2p^6}$ $\boxed{3s^2\,3p^6}$

From this we conclude that closed shells or sub-shells give very stable configurations.

(iv) *Transition Series Elements :* In the first three rows of the periodic table the properties (valency, ionisation energy, etc.) of the elements change uniformly from element to element. For the succeeding rows, however, this is not true. In the fourth row the elements from ^{21}Sc to ^{30}Zn have quite similar chemical properties. They form the first transition series. This observation can be easily explained. These elements occur during the filling of the *3d* sub-shell which is shielded by the already filled *4s* sub-shell from external influence. Hence the chemical properties of these elements, which depend on the electrons in the *outer* sub-shells of their atoms, are quite similar, independent of the number of their *3d* electrons. Similar transition series occur during the filling of *4d* and *5d* sub-shells in 5th and 6th periods.

(v) *Lanthanides and Actinides :* Lanthanides (or rare earths) are the elements from ^{58}Ce to ^{71} Lu. These are the elements in which the *4f* sub-shell is filling. This sub-shell lies within the outermost already filled *6s* sub-shell. Thus the *4f* electrons remain shielded from the external environment. Since the outer configuration, which is responsible for the chemical properties, is same for these elements, the chemical properties of the rare earths are almost identical. The same is true for the actinides, which occur during the filling of the $5f$ sub-shell inside the filled $7s$ sub-shell.

Importance of Exclusion Principle: If it were not obeyed, all the electrons in a multi-electron atom would be in the lowest-energy $1s$ sub-shell. In this case all atoms would have very high first excited states. Then all atoms would be extremely inert and there would be no molecules, no chemical compounds, no life. Infact, the world would be completely different if electrons did not obey the exclusion principle.

PROBLEMS

1. *If atoms could contain electrons with principal quantum number upto n = 6, how many elements would there be ?*

Solution : For a given n, the maximum number of electrons is $2n^2$

For $n = 1$,	max. no. of electrons	$= 2$
For $n = 2$,	"	$= 8$
For $n = 3$,	"	$= 18$
For $n = 4$	"	$= 32$
For $n = 5$,	"	$= 50$
For $n = 6$,	"	$= 72$
		182

Hence the number of elements would be 182.

2. *Use Pauli exclusion principle to show that the p sub-shell (l = 1) in an atom can contain a maximum of 6 electrons.*

Solution : No two electrons can have all the four quantum numbers n, l, m_l, m_s same. For electrons in the p sub-shell, n and l are same. Hence they must differ either in m_l or in m_s. For $l = 1$, we have

$$m_1 = 1, 0, -1$$

and we know $m_s = \pm \frac{1}{2}$.

The following different combinations of m_l and m_s, are possible :

m_l	1	0	−1	1	0	−1
m_s	$\frac{1}{2}$	$\frac{1}{2}$	$\frac{1}{2}$	$-\frac{1}{2}$	$-\frac{1}{2}$	$-\frac{1}{2}$

Thus a maximum of 6 electrons can occupy a p sub-shell.

3. ***Explain that 14 electrons will be required to completely fill the 4f sub-shell of an atom.***

Solution : For the 4f sub-shell, $n = 4$ and $l = 3$, so that

$$m_l = 3, 2, 1, 0, -1, -2, -3$$

and $$m_s = \pm \frac{1}{2}$$

We can see, as above, that 14 different combinations of m_l, and m_s, are possible; hence 14 electrons will complete an f-sub-shell.

4. ***Explain why an atomic shell with n = 2 cannot contain more than 8 electrons.***

Solution : According to Pauli's exclusion principle, no two electrons in an atom should have all the four quantum numbers n, l, j, m_j (or n, l. m_l, m,) same.

For $n = 2, l = 0, 1.$

For $l = 0, j = l \pm s = \frac{1}{2}$ and $m_j = \frac{1}{2}, -\frac{1}{2}$

and for $l = 1$, $j = \frac{3}{2}, \frac{1}{2}$

and $mj = \frac{3}{2}, \frac{1}{2}, -\frac{1}{2}, -\frac{3}{2}; \frac{1}{2}, -\frac{1}{2}$

Thus following different combinations of n, l, j, m_j are possible :

$$(2, 0, \frac{1}{2}, \frac{1}{2}) ; (2, 0, \frac{1}{2}, -\frac{1}{2}); (2, 1, \frac{3}{2}, \frac{3}{2}), (2, 1, \frac{3}{2}, \frac{1}{2});$$

$$(2, 1, \frac{3}{2}, -\frac{1}{2}); (2, 1, \frac{3}{2}, -\frac{3}{2}), (2, 1, \frac{1}{2}, \frac{1}{2}); (2, 1, \frac{1}{2}, -\frac{1}{2}).$$

Hence a maximum of 8 electrons can occupy $n = 2$ shell.

10

Raman Spectrum

Raman Effect

When a strong beam of visible or ultraviolet line-spectral light illuminates a gas, a liquid, or a transparent solid, a small fraction of light is scattered in all directions. The spectrum of the scattered light is found to consist of lines of the same frequencies as the incident beam (Rayleigh lines), and also certain weak lines of changed frequencies. These additional lines are called 'Raman lines'.

The Raman lines corresponding to each exciting (Rayleigh) line occur symmetrically on both sides of the exciting line. The lines on the low-frequency side of the exciting line are called 'Stokes' lines, while those on the high-frequency side are called 'anti-Stokes' lines. The anti-Stokes Raman lines are much weaker compared to the Stokes Raman lines. This phenomenon is called 'Raman effect'.

The displacements (in cm^{-1}) of the Raman lines from the corresponding exciting lines are independent of the frequencies of the latter. If another light source with a different line-spectrum is used, other Raman lines are obtained for the same scattering substance. However, the displacements from the exciting lines are the same. For different scattering substances, the displacements have different magnitudes. Thus *the Raman displacements are characteristic of the scattering substance.*

Experimental Setup: The basic requirements for photographing Raman spectrum are a source, a Raman tube and a spectrograph. The source must be an intense line-source with distinct lines in the blue-violet region. A mercury arc or a discharge lamp is a proper source. Now-a-days, laser provides an exceptionally intense and monochromatic Raman source.

The Raman tube used for liquids is a thin-walled glass (or quartz) tube *T* (Fig.) about 15 cm long and 2 cm in diameter, whose one end is closed with an optically-plane glass (or quartz) plate, and the other is drawn out into the shape of a horn and covered with black tape. The flat end serves as the window through which the scattered light emerges, while the blackened horn-shaped end causes the total reflection of the backward scattered light and provides a dark background.

The spectrograph must be one of high light-gathering power combined with good resolution. This may be achieved in a good prism spectrograph with a short-focus camera.

A typical arrangement for obtaining the Raman spectra of liquids is shown in Fig. The source *S* is a long horizontal arc of mercury. The Raman tube *T* containing the experimental liquid is placed above and parallel to the source *S*. In between the source and the tube is placed a glass cylindrical container filled with saturated solution of sodium nitrate. This acts as a cylindrical lens to concentrate the light along the axis of the Raman tube. The sodium nitrate solution absorbs the ultraviolet lines of the mercury arc but transmits the blue line with great intensity. A polished reflector *R* laid over *T* increases the intensity of illumination. The scattered light passing through the plane window of the Raman tube is focussed on the slit of a spectrograph which photographs the spectrum under a long exposure. A spectrophotometer may also be used as a recorder instead of a spectrograph.

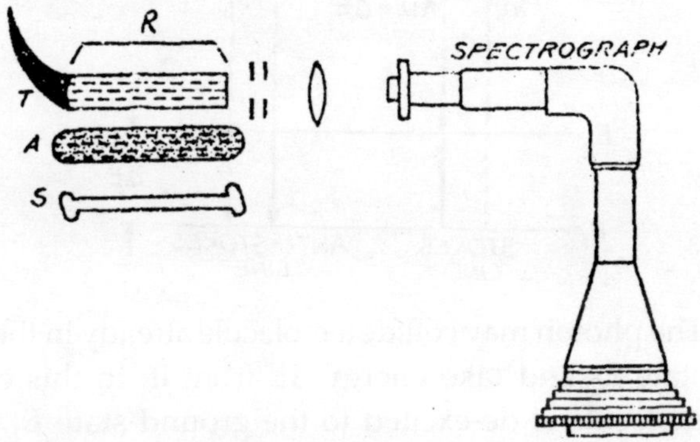

Explanation: Raman effect can be explained from quantum theory. According to this theory, light of frequency v is a bundle of 'photons', each of energy hv. When it falls on a scatterer, the photons collide with the molecules of the scatterer. There are three possibilities in such a collision :

(i) The photon may be scattered or deflected off its path without loss or gain of energy. It then gives rise to the unmodified spectral line of the same frequency v as of the incident light. This is the Rayleigh line.

(ii) The photon may give a part of its energy, ΔE (say), to a molecule which is in its ground energy state E_1 (Fig.). The molecule is then excited to a higher energy state $E_2 (= E_1 + \Delta E)$, and the photon is consequently scattered with a smaller energy $hv—\Delta E$. In this case it gives rise to a spectral line of lower frequency (or longer wavelength). This is the Stokes Raman line (Fig.).

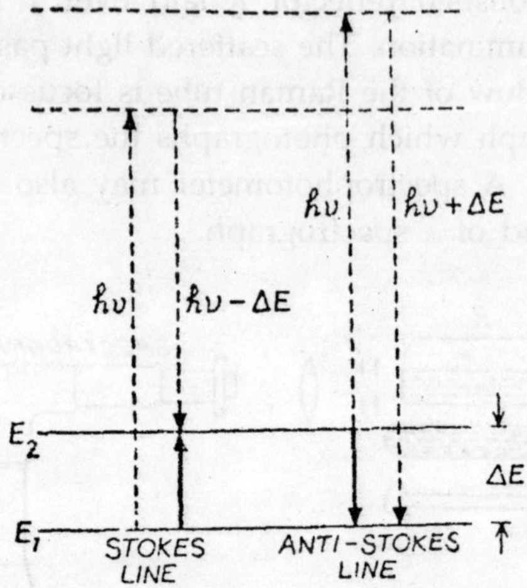

(iii) The photon may collide a molecule already in the excited state E_2 and take energy ΔE from it. In this case, the molecule is de-excited to the ground state E_1 and the photon is scattered with increased energy $hv + \Delta E$. Now it gives rise to a spectral line of higher frequency (or shorter wavelength). This is the anti-Stokes Raman line (Fig.).

Since the number of molecules in the excited state is very small, the chances of this last process are very small. Hence anti-Stokes Raman lines are much weaker than the Stokes Raman lines.

Uses : Raman effect is a powerful tool for studying the molecular structure of compounds and crystals. It is used to determine the arrangement of atoms in a given molecule. It supplies data regarding the spin and statistics of the nucleus. It is used in industries for studying the composition of mixtures, plastics, etc.

Compton Effect Versus Raman Effect: Both these phenomena are interaction of matter with radiation. In the Compton effect an electron *loosely* bound to an atom acquires energy from the incident photon and becomes free from the atom, the photon being scattered with reduced energy (or reduced frequency). The nucleus takes almost no part in the process. The Raman effect, on the other hand, is a very general case of interaction between photon and matter in which the entire molecule takes part. In this process the molecule acquires energy from the incident photon and is simply raised to an excited state of higher energy, no electron from the molecule is freed.

In case of Compton effect the wavelength of the scattered photon is always longer than that of the incident photon. In case of Raman effect, however, the scattered wavelength may be longer as well as shorter than the incident wavelength.

PROBLEMS

1. ***With exciting line 2536 Å a Raman line for a sample is observed at 2612 Å. Calculate the Raman shift in cm^{-1} units.***

Solution : The wave number $\bar{\upsilon}\left(=\frac{1}{\lambda}\right)$ of the exciting line is

$$\bar{\upsilon} = \frac{1}{2536 \times 10^{-8}\ \text{cm}} = 39432\ \text{cm}^{-1}$$

and that of the Raman line is

$$\bar{\upsilon}_{\text{Raman}} = \frac{1}{2612 \times 10^{-8}\ \text{cm}} = 38285\ \text{cm}^{-1}.$$

∴ Raman shift

$$\Delta\,\bar{\upsilon} = 39432–38285 = 1147\ \text{cm}^{-1}.$$

2. ***The exciting line in an experiment is 5460 Å and the Stokes line is at 5520 Å. Find the wavelength of the anti-Stokes line.***

Solution : The Stokes and anti-Stokes lines have the same wave-number displacement with respect to the exciting line. The wave-number of the exciting line is

$$v = \frac{1}{5460 \times 10^{-8}\ \text{cm}} = 18315\ \text{cm}^{-1}$$

and that of the Stokes line is

$$\frac{1}{5520 \times 10^{-8}\ \text{cm}} = 18116\ \text{cm}^{-1}.$$

Thus the wave-number displacement is

$$\Delta v = 18315\text{–}18116 = 199\ \text{cm}^{-1}.$$

Therefore, the wave-number corresponding to the anti-Stokes line would be given by

$$v + \Delta v = 18315 + 199 = 18514\ \text{cm}^{-1}.$$

The corresponding wavelength is

$$\frac{1}{18514\ \text{cm}^{-1}} = 5.401 \times 10^{-5}\ \text{cm} = 5401\ \text{Å}.$$

3. ***With exciting line 4358 Å, a sample gives Stokes line at 4458 A. Deduce the wavelength of the anti-Stokes line.***

[Ans. 4262 Å**]**

11

Zeeman Effect

Vector Model

Orbital Magnetic Dipole Moment of Atomic Electron: An electron revolving in an orbit about the nucleus of an atom is a minute current loop and produces a magnetic field. It thus behaves like a magnetic dipole. Let us calculate its magnetic moment.

Let us consider an electron of mass m and charge $-e$ moving with speed v in a circular Bohr orbit of radius r (Fig.), It constitutes a current of magnitude

$$i = \frac{e}{T},$$

where T is the orbital period of the electron.

Now, $$T = \frac{2\pi r}{v}, \text{ and so}$$

$$i = \frac{ev}{2\pi r}.$$

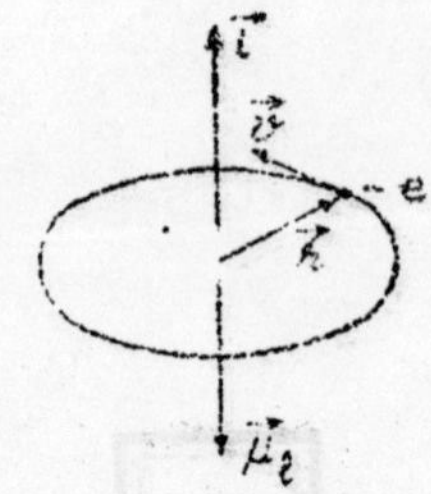

From electromagnetic theory, the magnitude of the orbital magnetic dipole moment $\vec{\mu}_l$ for a current i in a loop of area A is

$$\mu_l = iA$$

and its direction is perpendicular to the plane of the orbit as shown. Substituting the value of i from above and taking $A = \pi r^2$, we have

$$\mu_l = \frac{ev}{2\pi r}\pi r^2 = \frac{evr}{2} \qquad \text{... (i)}$$

Because the electron has a negative charge, its magnetic dipole moment $\vec{\mu}_l$ is opposite in direction to its orbital angular momentum $\vec{L}$, whose magnitude is given by

$$L = mvr. \qquad \text{... (ii)}$$

Dividing eq. (*i*) by eq. (*ii*), we get

$$\frac{\mu_l}{L} = \frac{e}{2m} \qquad \text{... (iii)}$$

Thus the ratio of the magnitude μ_l of the orbital magnetic dipole moment to the magnitude L of the orbital angular momentum for the electron is a constant, independent of the details of the orbit. This constant is called the 'gyromagnetic ratio' for the electron.

We can write eq. (*iii*) as a vector equation :

$$\boxed{\vec{\mu}_l = -\left(\frac{e}{2m}\right)\vec{L}.}$$

The minus sign means that $\vec{\mu}_l$ is in the opposite direction to $\vec{L}$.

The unit of electron magnetic moment is amp-m^2 or joule/tesla.

Bohr Magneton : From wave mechanics, the permitted scalar values of $\vec{L}$ are given by

$$L = \sqrt{[l(l+1)]}\frac{h}{2\pi},$$

where l is the 'orbital quantum number'. Therefore, the magnitude of the orbital magnetic moment of the election is

$$\mu_l = \sqrt{[l(l+1)]}\frac{eh}{4\pi m}.$$

The quantity $\frac{eh}{4\pi m}$ forms a natural unit for the measurement of atomic magnetic dipole moments, and is called the 'Bohr magneton', denoted by μ_B. Thus

$$\mu_l = \sqrt{[l(l+1)]}\,\mu_B,$$

where
$$\mu_B = \frac{eh}{4\pi m} = \frac{(1.6\times10^{-19}\text{ C})(6.63\times10^{-34}\text{ J-s})}{4\times3.14(9.1\times10^{31}\text{kg})}$$

$$= 9.28 \times 10^{-24}\text{ amp-m}^2.$$

Larmor Precession (Behaviour of a Magnetic Dipole in an External Magnetic Field) : An electron moving around the nucleus of an atom is equivalent to a magnetic dipole. Hence, when the atom is placed in an external magnetic field, the electron orbit precesses about the field direction as axis. This precession is called the 'Larmor precession, and the frequency of this precession is called the 'Larmor frequency'.

In Fig. is shown an electron orbit in an external magnetic field $\vec{B}$. The orbital angular momentum of the electron is represented by a vector $\vec{L}$ perpendicular to the plane of the orbit. Let θ be the angle between $\vec{L}$ and $\vec{B}$.

The orbital dipole moment $\vec{\mu}_l$ of the electron is given by

$$\vec{\mu}_l = -\left(\frac{e}{2m}\right)\vec{L}, \qquad \text{... (i)}$$

where $-e$ is the charge on the electron of mass m. The minus sign signifies that $\vec{\mu}_l$ is directed opposite to $\vec{L}$. As a result of its interaction with $\vec{B}$, the dipole experiences a torque $\vec{\tau}$, given by

$$\vec{\tau} = \vec{\mu}_l \times \vec{B}. \qquad \text{... (ii)}$$

According to eq. (*i*) and (*ii*), the torque $\vec{\tau}$ is always *perpendicular* to the angular momentum $\vec{L}$.

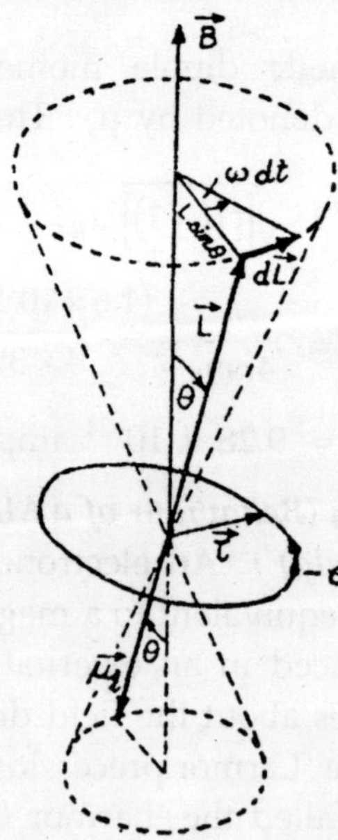

We know that a torque causes the angular momentum to change according to a form of Newton's law

$$\vec{\tau} = \frac{d\vec{L}}{dt},$$

and the change takes place in the direction of the torque. The torque $\vec{\tau}$ on the electron, therefore, produces a changed $\vec{L}$ in $\vec{L}$ in a time dt. *The changed* $\vec{L}$ *is perpendicular to* $\vec{L}$ (because the change is in the direction of torque, and the torque is perpendicular to $\vec{L}$). Hence the angular momentum $\vec{L}$ remains constant in magnitude, but its direction changes. As time goes on, $\vec{L}$ traces a cone around $\vec{B}$, such that the angle between $\vec{L}$ and $\vec{B}$ remains constant. This is the precession of $\vec{L}$, and hence of the electron orbit, around $\vec{B}$.

If ω be the angular velocity of precession, then $\vec{L}$ precesses through an angle $\omega\, dt$ in time dt. From the Fig., we see that

$$\omega\, dt = \frac{dL}{L \sin\theta} \qquad \left[\text{angle} = \frac{\text{arc}}{\text{radius}}\right]$$

or

$$\omega = \frac{dL}{dt}\frac{1}{L\sin\theta} = \frac{\tau}{L\sin\theta}.$$

But. from eq. (*ii*), $\tau = \mu_l\, B \sin\theta$.

$$\therefore \qquad \omega = \frac{\mu_l}{L} B$$

Thus the angular velocity of Larmor precession is equal to the product of the magnitude of the magnetic field and the ratio of the magnitude of the magnetic moment to the magnitude of the angular momentum.

Again, from eq. (*i*), $\dfrac{\mu_l}{L} = \dfrac{e}{2m}$.

$$\therefore \qquad \omega = \frac{e}{2m} B.$$

The Larmor frequency (frequency of precession) is therefore

$$f = \frac{\omega}{2\pi} = \frac{e}{4\pi m}B.$$

It is independent of the orientation angle θ between orbit normal ($\vec{L}$) and field direction ($\vec{B}$).

Importance: This theorem is of considerable importance in atomic structure as it enables an easy calculation of energy levels in the presence of an external magnetic field.

Nr : — Putting $e = 1.6 \times 10^{-19}$ coul, $m = 9.1 \times 10^{-31}$ kg and B — 10^4 weber/m^2 (given), we get

$$f = 1.4 \times 10^{14} \text{ per second.}$$

Space Quantisation of Atoms: An electron moving around the nucleus of an atom is equivalent to a magnetic dipole. Hence, when the atom is placed in an external magnetic field $\vec{B}$, the electron orbit precesses about the field direction as axis. The electron orbital angular momentum vector $\vec{L}$ traces a cone around $\vec{B}$ such that the angle θ between $\vec{L}$ and $\vec{B}$ remains constant (Fig.).

If the magnetic field $\vec{B}$ is along the z-axis, the component of $\vec{L}$ parallel to the field is

$$L_z = L \cos\theta$$

or $$\cos\theta = \frac{L_z}{L}$$

Quantum mechanically, the angular momentum L and its z-component L_z are quantised according to the relations

$$L = \sqrt{[l(l+1)]}\,\frac{h}{2\pi}$$

and $$L_z = m_l \frac{h}{2\pi},$$

where l and m_l are the orbital and magnetic quantum numbers respectively. Therefore,

$$\cos\theta = \frac{L_z}{L} = \frac{m_l}{\sqrt{[l(l+1)]}}.$$

Thus the angle θ between $\vec{L}$ and the z-axis is determined by the quantum numbers l and m_l. Since, for a given l, there are $(2l+1)$ possible values of ml (= 0, ± 1, ± 2,........, ± 1), the angle θ can assume $(2l + 1)$ discrete values. In other words, the angular momentum vector $\vec{L}$ can have $(2l+1)$ discrete orientations with respect to the magnetic field. This quantisation of the orientation of atoms in space is known as 'space quantisation'.

The space quantisation of the orbital angular momentum vector $\vec{L}$ corresponding to $l = 2$ {or $L = \sqrt{(6)}\, h/2\pi$} is shown in Fig. For $l = 2$, we have

$$ml = 2, 1, 0, -1, -2$$

so that

$$L_z = 2\frac{h}{2\pi}, \frac{h}{2\pi}, 0, -\frac{h}{2\pi}, -2\frac{h}{2\pi}.$$

Alternatively, the orientations θ of $\vec{L}$ with respect to the field $\vec{B}$ (*z-axis*) are given by

$$\cos\theta = \frac{m_l}{\sqrt{[l(l+1)]}}$$

$$= \frac{2}{\sqrt{(6)}}, \frac{1}{\sqrt{(6)}}, 0, -\frac{1}{\sqrt{(6)}}, -\frac{2}{\sqrt{(6)}},$$

$$= 0.8165, 0.4082, 0, -0.4082, -0.8165$$

or

$$\theta = 35^\circ, 66^\circ, 90^\circ, 114^\circ, 145^\circ, 190^\circ.$$

We note that $\vec{L}$ can never be aligned exactly parallel or antiparallel to $\vec{B}$, since $|m_l|$ is always smaller than $\sqrt{[l\,(l+1)]}$.

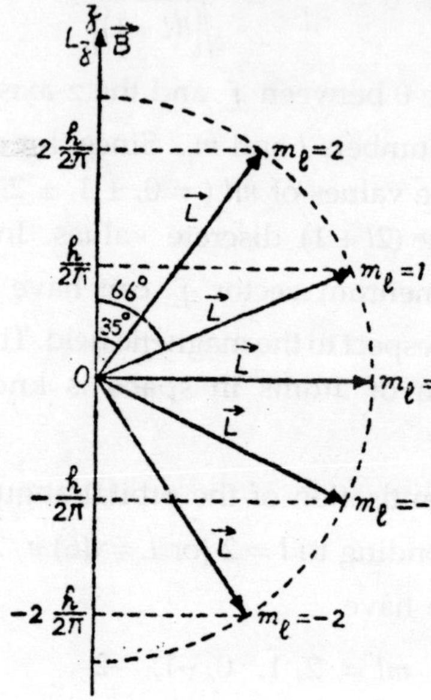

Zeeman Effect

Zeeman, in 1896, observed that when a light-source giving line spectrum is placed in an *external* magnetic field, the spectral lines emitted by the atoms of the source are split into a number of polarised components. This effect of magnetic field on the atomic spectral lines is called 'Zeeman effect'.

To produce Zeeman effect, a source of light, such as a gas discharge tube, is placed symmetrically between the pole-pieces of a strong electromagnet, one of whose pieces carries a hole drilled parallel to the magnetic field direction. The light coming from the tube is examined by a spectroscope of high resolving power. The attention is focused on a single spectral line observable in the absence of the magnetic field (Fig.).

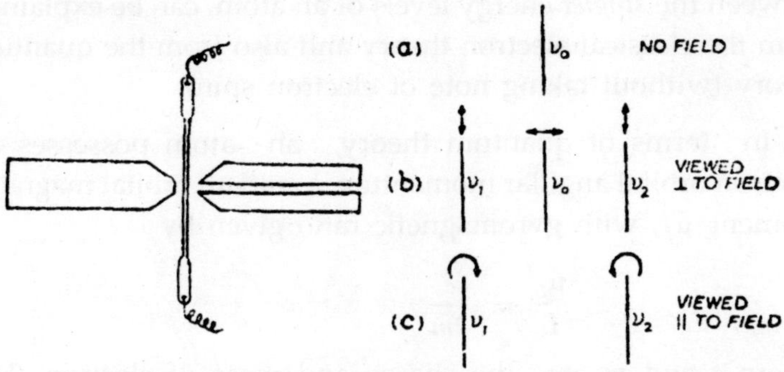

When the light is viewed at right angles to the magnetic field direction, a singlet spectral line is found to split up into three components (Fig.). The central component is in the same position and hence has the same frequency v_0 as the original line. The outer components of frequencies v_1 and v_2 are displaced equally from the central component. The central component is linearly polarised with electric vector vibrating parallel to the magnetic field, while the two outer components are linearly polarised with electric vector vibrating at right angles to the field. This splitting is known as *'normal* transverse Zeeman effect'.

When the light is viewed (through the hole) in a direction parallel to the magnetic field, only the two outer components are seen and there is no central component (Fig.). These components are circularly polarised in opposite senses. This is known as *'normal* longitudinal Zeeman effect'.

The fine-structure components of a *multiplet* spectral line, however, show a complex Zeeman pattern. For example, the D_1 and D_2 components of sodium yellow doublet give four and six lines respectively in the Zeeman pattern. This is 'anomalous' Zeeman effect.

Explanation of Normal Zeeman Effect : The normal Zeeman effect, which is shown by spectral lines arising from transitions between the *singlet* energy levels of an atom, can be explained from the classical electron theory and also from the quantum theory (without taking note of electron spin).

In terms of quantum theory, an atom possesses an electron orbital angular momentum $\vec{L}$ and an orbital magnetic moment $\vec{\mu}_l$, with gyromagnetic ratio given by

$$\frac{\mu_l}{L} = \frac{e}{2m},$$

where e and m are the charge and mass of electron. The vector $\vec{\mu}_l$ is directed opposite to the vector $\vec{L}$ because the electron is negatively charged (Fig.).

When the atom is placed in an external magnetic field $\vec{B}$ say along the z axis, the vector $\vec{L}$ precesses around the field direction (Larmor precession) with quantised components given by

$$L_z = m_l \frac{h}{2\pi},$$

where m_l is the magnetic quantum number.

By Larmor's theorem, the angular velocity of precession is given by

$$\omega = \frac{\mu_l}{L} B = \frac{e}{2m} B.$$

The change in energy of such a precession is equal to the product of the angular velocity and the component of angular momentum along the field, that is,

$$\Delta E = \omega L_z = \frac{eB}{2m} mt \frac{h}{2\pi} = \text{mt}\ \frac{ehB}{4\pi m}. \quad \dots (i)$$

Now, *ml* takes discrete values from + l to —l, *i.e.*

mt=ml =l, l–1,...2, 1,0, – 1, –2, ...l, a total of (2l+1) values. Thus each energy level (of a given l value) of the atom placed in the magnetic field is split into (2l + 1) energy levels, called

'Zeeman levels', having different m_l, values. The separation between the Zeeman levels is same for all the energy levels of the atom.

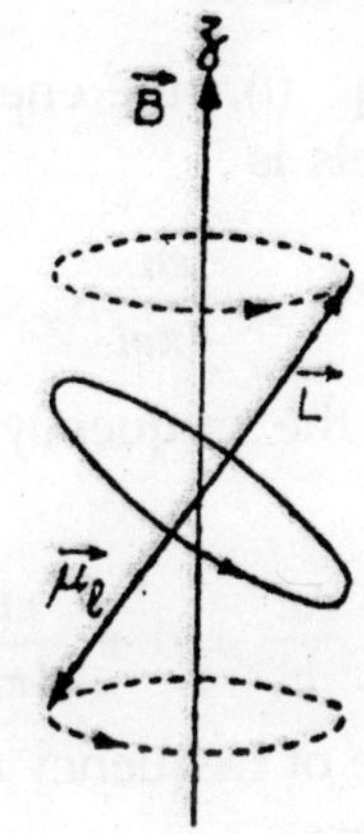

Let us consider a line arising by electron transition from $l = 1$ level to $l = 0$ level (Fig.). In the magnetic field, the level $l = 1$ is split into $(2l + 1) = 3$ components corresponding to $ml = +1, 0, -1$; while the level $l = 0$ remains unplatted (Fig.).

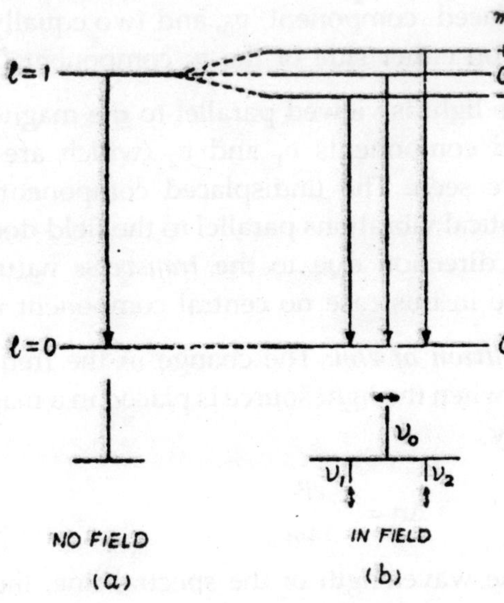

Theory shows that only those transitions are allowed for which the quantum number m_l changes by 0 or ± 1 , *i.e.*

$$\Delta m_l = 0 \pm 1.$$

Therefore, from eq. (*i*), the energy-change due to the Zeeman splitting of levels is

$$\Delta E = 0, \pm \frac{eh}{4\pi m} B.$$

Hence the change in the frequency of the emitted spectral line is given by

$$\Delta \nu = \frac{\Delta E}{h} = 0 \pm \frac{eB}{4\pi m}.$$

Thus the original line of frequency ν_0 is splitted into three components of frequencies

$$\nu_1 = \nu_0 + \frac{eB}{4\pi m},$$

and
$$\nu_2 = \nu_0 - \frac{eB}{4\pi m},$$

i.e. one displaced component ν_0, and two equally-displaced components on either side of the ν_0 component (Fig.).

When the light is viewed parallel to the magnetic field B, the displaced components ν_1 and ν_2 (which are circularly-polarised) are seen. The undisplaced component, however, which has optical vibrations parallel to the field does not send light in this direction due to the *transverse* nature of light waves. Hence in this case no central component is seen.

Determination of e/m: The change in the frequency of a spectral line, when the light source is placed in a magnetic field B, is given by

$$\Delta \nu = \frac{eB}{4\pi m}$$

If λ be the wavelength of the spectral line, then

$$v = \frac{c}{\lambda}$$

or $$\Delta v = -\frac{c}{\lambda^2}\,\Delta\lambda$$

$$\therefore \qquad \Delta\lambda = -\frac{\lambda^2}{c}\,\Delta v = -\frac{\lambda^2}{c}\frac{eB}{4\pi m}$$

or $$\Delta\lambda = \frac{eB\lambda^2}{4\pi mc} \text{ (numerically).}$$

$$\therefore \qquad \frac{e}{m} = \frac{4\pi c}{B}\left(\frac{\Delta\lambda}{\lambda^2}\right).$$

Hence, measuring the wavelength-change, e/m can be evaluated.

Electron Spin

The Bohr-Sommerfeld quantum theory of elliptic orbits *with relativity correction* was in fair agreement with the observed fine structure of hydrogen spectral lines. It, however, suffered from two major drawbacks :

Firstly, it could not explain the fine-structure observed in the spectral lines of atoms other than hydrogen. For example, the spectral lines of alkali atoms are *doublets,* have two close fine-structure components. In alkali atoms, the (single) optical electron moves in a Bohr-like orbit of *large* radius at *low* velocity. Therefore, the relativity effect would be too small to account for the large fine-structure splitting observed in alkali lines.

Secondly, the simple quantum theory failed to explain anomalous Zeeman effect, that is, the splitting of atomic spectral lines into four, six, or more, components when the light source was placed in an external magnetic field.

In view of these drawbacks of the theory, Goudsmit and Uhlenbeck proposed in 1925 that *an electron must be looked upon as a charged sphere spinning about its own axis having an intrinsic*

angular momentum and an intrinsic magnetic moment. These are called 'spin angular momentum' $\vec{S}$ and 'spin magnetic moment' $\vec{\mu}_s$, respectively. (These are besides the orbital angular momentum $\vec{L}$ and orbital magnetic moment $\vec{\mu}_l$). The magnitude of the spin angular momentum is

$$S = \sqrt{[s(s+1)]}\frac{h}{2\pi},$$

where s is the 'spin quantum number'. The only value s can have is

$$S = \frac{1}{2},$$

as required by experimental data. Thus

$$S = \sqrt{\left[\frac{1}{2}\left(\frac{1}{2}+1\right)\right]}\frac{n}{2\pi} = \frac{\sqrt{3}}{2}\frac{h}{2\pi}.$$

The component of $\vec{S}$ along a magnetic field parallel to the z-direction is

$$S_z = m_s \frac{h}{2\pi},$$

where m_s is the 'spin magnetic quantum number' and takes $(2s + 1) = 2$ values which are $+ s$ and $- s$, that is

$$m_s = +\frac{1}{2} \text{ and } -\frac{1}{2}.$$

Thus $$s_z = +\frac{1}{2}\frac{h}{2\pi} \text{ and } = -\frac{1}{2}\frac{h}{2\pi}$$

The gyromagnetic ratio for electron spin, $\frac{\mu_s}{S}$ is *twice* the corresponding ratio $\frac{\mu_l}{L}\left(=\frac{e}{2m}\right)$ for the electron orbital motion. Thus the spin magnetic moment $\vec{\mu}_s$ of electron is related to the spin angular momentum $\vec{S}$ by

$$\vec{\mu}_s = -\frac{e}{m}\vec{S}.$$

The minus sign indicates that $\vec{\mu}_s$ is opposite in direction to $\vec{S}$ (because electron is negatively charged).

The magnitude of the spin magnetic moment is

$$\mu_s = \frac{e}{m}S.$$

$$= \frac{e}{m}\cdot\frac{\sqrt{3}}{2}\frac{h}{2\pi}$$

$$= \sqrt{3}\frac{eh}{4\pi m}$$

$$= \sqrt{3}\,\mu_{B},$$

where μ_B is the Bohr magneton.

Coupling of Orbital and Spin Angular Momenta : Vector Model of the Atom: The total angular momentum of an atom results from the combination of the orbital and spin angular momenta of its electrons. Since angular momentum is a vector quantity, we can represent the total angular momentum by means of a vector, obtained by the addition of orbital and spin angular momentum vectors. This leads to the vector model of the atom.

Let us consider an atom whose total angular momentum is provided by a single electron. The magnitude of the orbital angular momentum $\vec{L}$ of an atomic electron is given by

$$L = \sqrt{[l(l+1)]}\frac{h}{2\pi}$$

and its z component

$$Lz = m_l\frac{h}{2\pi},$$

where l is orbital quantum number and ml, is the corresponding magnetic quantum number, with values

$$m_l = l, l - 1, \ldots\ldots, - l + 1, - l.$$

Similarly, the magnitude of the spin angular momentum $\vec{S}$ is given by

$$S = \sqrt{s(s+1)}\,\frac{h}{2\pi}$$

and its z-component

$$S_z = m_s \frac{h}{2\pi},$$

where s is the spin quantum number (which has the sole value $+\frac{1}{2}$) and m_s is magnetic spin quantum number ($m_s = \pm\frac{1}{2} = \pm s$).

The total angular momentum of the one-electron atom, $\vec{J}$,is the vector sum of $\vec{L}$ and $\vec{S}$, that is

$$\vec{J} = \vec{L} + \vec{S}.$$

The magnitude and the z-component of $\vec{J}$ are specified by two quantum numbers j and m_j, according to the usual quantisation conditions

$$J = \sqrt{j(j+1)}\,\frac{h}{2\pi}$$

and $$J_z = mj \frac{h}{2\pi}.$$

j is called the 'inner quantum number' and m_j is the corresponding magnetic quantum number. The possible values of m, range from $+ j$ to $- j$ in integral steps :

$$m_j = j, j - 1\ldots\ldots, - j + 1, -j.$$

Let us obtain the relationship among the various angular momentum quantum numbers. Since J_z, L_x and S_z are scalar quantities, we may write

$$J_z = L_z \pm S_z.$$

This gives $\quad m_j = mi \pm m_s.$

The maximum values of m_1, m_l and m_s are j, l and s respectively. Therefore, we have

$$\boxed{j = l \pm s}$$

Since $\vec{J}$, $\vec{L}$ and $\vec{S}$ are ail quantised, they can have only certain specific relative orientations. In case of a one-electron atom, there are only two relative orientations possible, corresponding to

$$j = l + s,$$

so that $\quad J > L.$

and $\quad j = l - s,$

so that $\quad J < L.$

The two ways in which $\vec{L}$ and $\vec{S}$ can combine to form $\vec{J}$ (when $l = 1$, $s = \frac{1}{2}$) are shown in Fig.

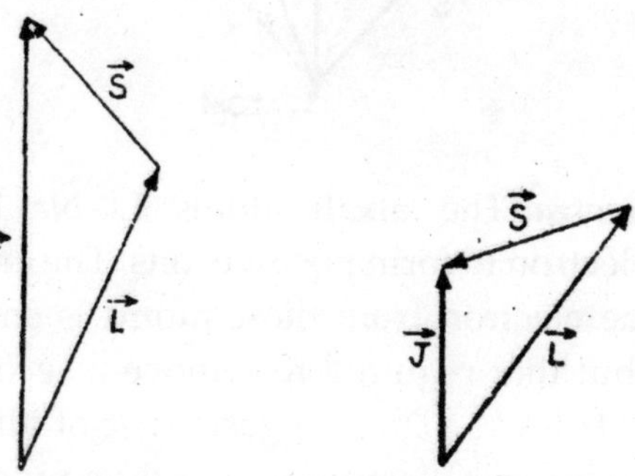

$j = l + s = \frac{3}{2}$ $\qquad$ $j = l - s = \frac{1}{2}$

The angular momenta of the atomic electron, $\vec{L}$ and $\vec{S}$, interact magnetically; which is known as 'spin-orbit interaction'. They exert torques on each other. These *internal* torques do not change the magnitudes of the vectors $\vec{L}$ and $\vec{S}$, but cause them to precess uniformly around their resultant $\vec{J}$ (Fig.). If the atom is in free space so that no external torques act on it, then the total angular momentum $\vec{J}$ is conserved in magnitude and direction. Obviously, the angle between $\vec{L}$ and $\vec{S}$ would remain invariant. This is the vector model of the one-electron atom. It can be extended to many-electron atoms. The vector model enables us to explain the phenomena which could not be understood from Bohr-Sommerfeld theory such as fine structure of spectral lines and anomalous Zeeman effect.

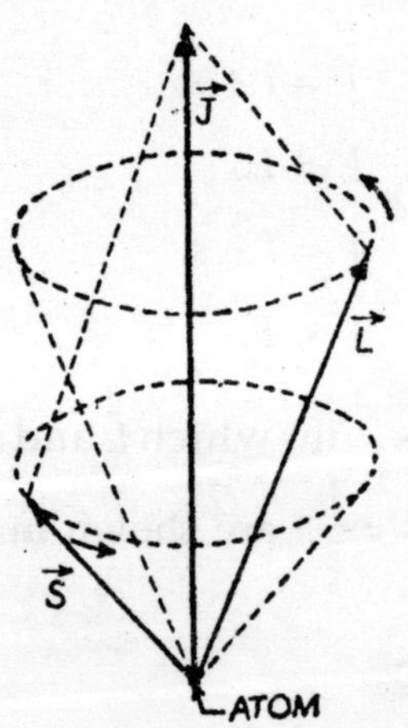

Alkali Spectra: The alkali atoms Li, Na, K,.........readily give up one electron to form positive ions. The energy required to remove one electron from these atoms is small (5.1 eV in case of Na), but that required to remove a second electron is much larger (47.3 eV). This suggests that of all the electrons in an alkali atom, *one* electron is loosely-bound to the atom. The spectral lines of an alkali atom arise due to the transitions of this electron only which is called the 'optical' or 'valence' electron. The alkali spectrum is called 'one-electron' spectrum.

The alkali spectrum consists of spectral lines which can be classified into four series : principal, series, sharp series, diffuse series and fundamental series. The principal series is the most prominent and can be observed in emission as well as in absorption spectrum. The other series are observed in emission spectrum only.

The emission of alkali spectral lines can be fairly explained on the same lines as the Bohr-Sommerfeld theory for hydrogen atom. An atom has a number of discrete energy states, each state being characterised by a total quantum number $n(=1, 2, 3, \ldots \infty)$. For each value of n, there are component levels labelled by an additional quantum number l, called the 'orbital' quantum number, l can take values 0, 1, 2,......$(n - 1)$. Thus $n = 1$ state has only one level $(l = 0)$; $n = 2$ state has two levels $(l = 0, 1)$, and so on. The levels corresponding to $l = 0, 1, 2, 3, \ldots$ are called as *s, p, d, f*,....levels respectively. Thus $n - 1$ state has a level called 1s; the $n = 2$ state has two levels called 2*s* and 2*p* ; the $n = 3$ state has three levels called 3*s*, 3*p* and 3*d*; the $n = 4$ state has four levels 4*s*, 4*p*, 4*d*, 4*f*, and so on. The energies of these levels are given by

$$E_{n,l} = -\frac{Rhc}{(n-\Delta)^2},$$

where Δ is called 'quantum defect' and depends on l. Thus the energies of levels with same n but different l are *different*.

Let us now consider an alkali atom, say sodium (Na). Measurement of spectral wavelengths and the ionisation potential shows that the ground state of the sodium atom is 3*s*, *i.e.* the optical electron occupies the 3*s* level in the normal atom. When the atom is excited by some outer means, the electron leaves the 3*s* level and goes to *any* of the higher levels 3*p*, 3*d*, 4*s*, 4*p*, 4*d*, 4*f*,.... From the higher level the electron jumps back to the lower levels but only such that the l value changes by ± 1 (Fig.), *i.e.*

$$\Delta l = \pm 1.$$

This is called the 'selection rule'. When the electron jumps from any *p*-level to the lowest *s*-level (3*s*), it emits a line of principal series ; when it jumps from any *s*-level to the lowest *p*-level (3*p*), it emits a line of sharp series ; and so on. Thus emission of all the spectral series is explained.

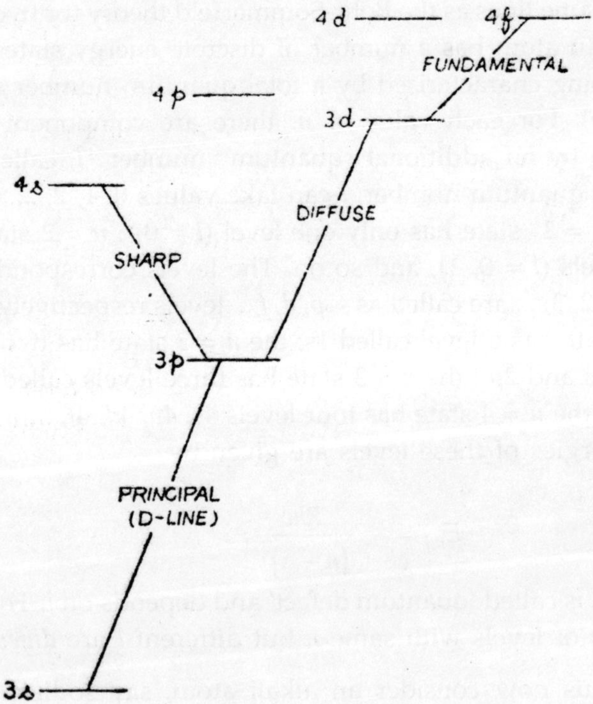

Fine Structure of Alkali Spectra and its Explanation: When the spectral lines of an alkali atom are seen under high resolution, each of them is found to consist of two close components. For example, the yellow *D*-line of sodium consists of two close lines of wavelengths 5890 Å and 5896 Å. This is called the fine structure and is explained by introducing the conception of electron spin.

The total angular momentum of the electron is the *vector* sum of its orbital angular momentum $\sqrt{[l(l+1)]}\,\frac{h}{2\pi}$ and the spin angular momentum $\sqrt{[s(s+1)]}\,\frac{h}{2\pi}$ and is given by $\sqrt{[j(j+1)]}\,\frac{h}{2\pi}$, where j is called the 'inner quantum number'. It takes the values given by

$$j = l \pm s = l \pm \frac{1}{2}.$$

Thus, for

$$l = 0 \text{ (s-level)}, j = \frac{1}{2} \quad (j \text{ cannot be negative})$$

$$l = 1 \text{ (p-level)}, j = \frac{3}{2}, \frac{1}{2}$$

$$l = 2 \text{ (d-level)}, j = \frac{5}{2}, \frac{3}{2}, \text{ and so on.}$$

Each value of the total angular momentum of the electron corresponds to a particular total energy of the electron. Therefore, each energy level of a given l-value is split into two sub-levels of slightly different energies, corresponding to the two j-values. The s-levels (for which $l = 0$) still remain unsplitted because there is only one j-value for them. On this basis, the doublet structure of the spectral lines of alkali atoms can be explained.

Let us consider the sodium D-line. It arises from the transition of the electron from the $3p$ to the $3s$ level (Fig.). But, due to electron spin, the $3p$ level consists of two sub-levels, one corresponding to $j = \frac{3}{2}$ and the other corresponding *to* $j = \frac{1}{2}$ (Fig.). Thus there are two transitions and hence the D-line is a doublet.

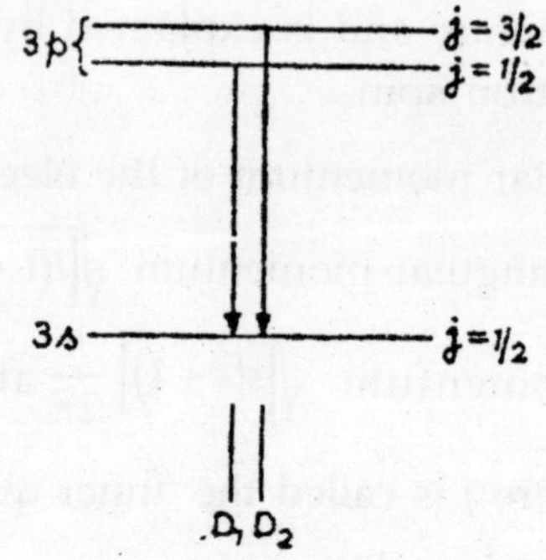

D-LINE DOUBLET

Similarly, the spin-orbit coupling together with relativity correction explains the fine structure of hydrogen lines.

Stern-Gerlach Experiment : Stern and Gerlach, in 1921, performed an experiment which demonstrated directly that an atom placed in a magnetic field can take only certain discrete orientations with respect to the field (space quantisation). It also demonstrated the existence of electron spin, thus providing an experimental verification of the vector model of atom.

The plan of the experiment is shown in Fig. A beam of neutral silver atoms was formed by heating silver in an oven. It was collimated by a few fine slits and then passed through a *non-homogeneous* magnetic field. The field was produced by specially-designed pole-pieces whose cross-sectional view is displayed separately. It shows that the field increases in intensity in the z-direction defined in the figure.

The beam leaving the magnetic field was received on a photographic plate *P*. On developing the plate, no trace of the direct beam was obtained. Instead, two traces were obtained, symmetrically situated with respect to the direct beam. This meant that the beam of silver atoms splitted into two discrete components, one component being bent in the + z-direction and the other bent in the — z-direction. The experiment was repeated using other atoms, and in each case the beam was found splitted into two, or more, discrete components. This result is interpreted in the following way :

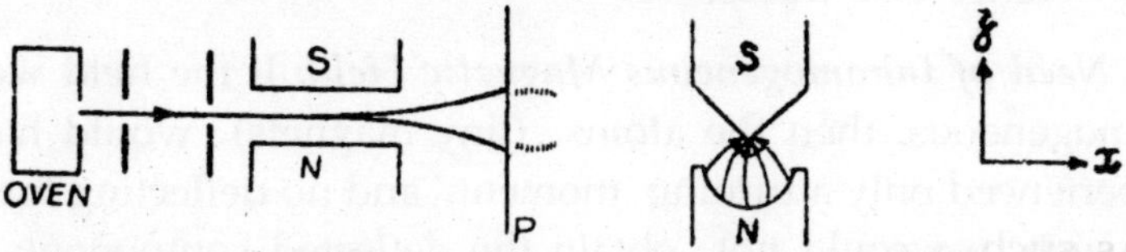

A magnet experiences a net *deflecting* force in a non-homogeneous magnetic field which depends on the orientation of the magnet in the field. Since atoms are tiny magnets, they experience deflecting force when passing through the field. If an atom could have *any* orientation in the magnetic field, then for the millions of atoms present in the beam, all possible orientations would be obtained and the beam would be deflected into a continuous band.

In the experiment, however, there was no band, but discrete traces on the photographic plate. This showed that *the atoms passing through the field were oriented in space in discrete directions* so that the beam deflected in certain discrete directions only and gave discrete traces on the plate.

The experiment is also an evidence for the existence of electron spin. This was shown most clearly in 1927 by Phipps and Taylor, who repeated the Stern Gerlach experiment by using a beam of hydrogen atoms. This atom consists of a single electron which, in the ground state, lies in an *s*-level, for which the quantum number $l = 0$. If there were no spin, then j would also be zero, so that $(2j + .1) = 1$.

In that case the hydrogen atomic beam would be unaffected by the magnetic field, and only one trace would be obtained on the plate. Phipps and Taylor, however, found the beam to be splitted into two symmetrically deflected components giving rise to two traces. This is just the case when the existence of electron spin is admitted and a value $\frac{1}{2}$ is assigned to the spin quantum number. Thus $j = l \pm s = 0 \pm \frac{1}{2} = \frac{1}{2}$ so that $2j + 1 = 2$. Hence two traces.

Need of Inhomogeneous Magnetic Field: If the field were homogeneous, then the atoms (tiny magnets) would have experienced only a turning moment, and no deflecting force. As such, we could not obtain the deflected components in spite of the orientation of the atoms relative to the field.

Atoms, not Ions: In the Stern-Gerlach experiment, a beam of 'neutral' atoms is passed through an inhomogeneous magnetic field, and each atom experiences a transverse force depending upon its orientation with respect to the field. If (charged) ions were used, they would be subjected to Lorentz force also and their deflection would no longer be transverse so that no traces would be obtained on the plate.

PROBLEMS

1. ***Calculate the two possible orientations of spin vector*** $\vec{S}$ ***with respect to magnetic field direction.***

Solution : Let the magnetic field $\vec{B}$ be along the z-axis. The magnitude of the spin angular momentum $\vec{S}$ and its z-component are quantised according to the relations

$$S = \sqrt{[s(s+1)]}\,\frac{h}{2\pi}, \qquad s = \frac{1}{2}$$

and

$$S_z = m_s \frac{h}{2\pi}, \qquad m_s = \pm\frac{1}{2}$$

Hence the angle θ between $\vec{S}$ and the z-axis (Fig.) is determined by the quantum numbers m_s *and* s, according as

$$\cos\theta = \frac{S_z}{S} = \frac{m_s}{\sqrt{[s(s+1)]}}$$

$$= \frac{2}{\sqrt{3}}\, m_s \qquad \left[\because s = \frac{1}{2}\right]$$

For $\quad m_s = +\frac{1}{2}$, we have

$$\cos\theta = +\frac{1}{\sqrt{3}} = 0.577.$$

$$\therefore \quad \theta = \cos^{-1}(0.577) = 114^\circ,$$

For $\quad m_s = \frac{1}{2}$, we have

$$\cos\theta = -\frac{1}{\sqrt{3}} = -0.577.$$

$$\therefore \quad \theta = \cos^{-1}(-0.557) = 125.3^\circ.$$

These angles are indicated in the diagram.

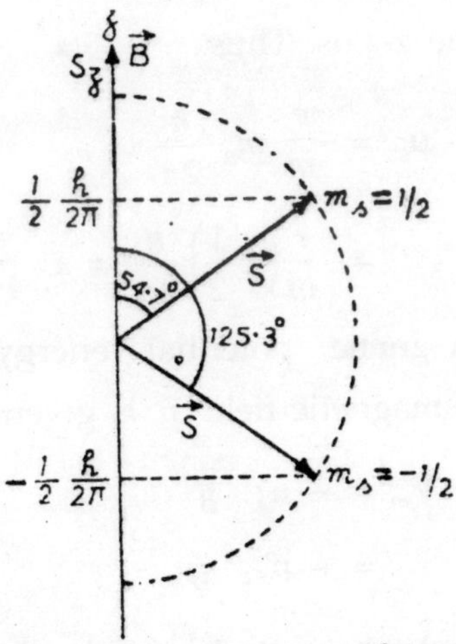

2. *A beam of electrons enters a uniform magnetic field of 1.2 Tesla. Calculate the energy difference between electrons whose spins are parallel and anti-parallel to the field.*

Solution : An electron has an intrinsic (spin) angular momentum $\vec{S}$ and an intrinsic magnetic dipole moment $\vec{\mu}_s$, which are related by

$$\vec{\mu}_s = -\frac{e}{m}\vec{S},$$

where $$S = \sqrt{[s(s+1)]}\frac{h}{2\pi} \text{ and } s = \frac{1}{2}.$$

Let the magnetic field $\vec{B}$ be along the z-axis. The magnitude of the z-component of the magnetic moment is

$$\mu_x = \frac{e}{m} S_z$$

where S_z is the z-component of the spin angular momentum and is given by

$$S_z = m_s \frac{h}{2\pi},$$

where $m_s = \pm\frac{1}{2}$, depending upon whether S_z is parallel or anti-parallel to the z-axis. Thus

$$\mu_{sz} = \frac{e}{m} m_s \frac{h}{2\pi}$$

$$= \frac{e}{m}\left(\pm\frac{1}{2}\right)\frac{h}{2\pi} = \pm\frac{eh}{4\pi m}.$$

Now, the magnetic potential energy of a dipole of moment $\vec{\mu}_s$, in a magnetic field $\vec{B}$ is given by

$$V_m = -\vec{\mu}_s \cdot \vec{B}$$

$$= -\mu_{sz}\, \vec{B},$$

where μ_{sz} is the scalar magnitude of $\vec{\mu}_s$ in the direction of the magnetic field. Thus

$$V_m = \pm\frac{eh}{4\pi m} B.$$

The difference in energy of the electrons having spin parallel and anti-parallel to the field is

$$\Delta V_m = \frac{ehB}{4\pi m} - \left(-\frac{ehB}{4\pi m}\right) = \frac{ehB}{2\pi m}.$$

Substituting the known values of *e*, *h* and *m* ; and the given values of *B*, we get

$$\Delta V_{\text{m}} = \frac{\left(1.6 \times 10^{-19} \text{ coul}\right) \times \left(6.63 \times 10^{-34} \text{ joule-sec}\right) \times \left(1.2 \text{ nt / amp} - \text{m}\right)}{2 \times 3.14 \times \left(9.1 \times 10^{-31} \text{ kg}\right)}$$

$= 2.23 \times 10^{-23}$ joule.

But $\quad 1\text{eV} = 1.6 \times 10^{-19}$ joule.

$$\therefore \quad \Delta V_{\text{m}} = \frac{2.23 \times 10^{-23}}{1.6 \times 10^{-19}} = 1.39 \times 10^{-4} \text{ eV}.$$

3. *(a) For a one-electron atom, write down the spectroscopic symbols for the possible energy levels of an electron with l = 2. (b) Which of these levels has the higher energy and why? (c) If the atom is placed in a weak magnetic field, into how many magnetic levels will each of the above levels split up? (d) Which one of these magnetic levels will have the highest energy and why ?*

Solution : (*a*) For the given election, we have

$$l = 2 \text{ and } s = \frac{1}{2}.$$

Therefore, the possible values of inner quantum number *j* are

$$j = l \pm s = 2 \pm \frac{1}{2} = \frac{5}{2}, \frac{3}{2}.$$

The multiplicity of the energy levels, defined by 2s + 1, is 2.

The energy levels for the election are designated by *S, P, D*,......according as $l = 0, 1, 2$,...... The multiplicity is indicated by a super-script and the value of *j* by a sub-script. Thus the possible energy levels for $l = 2$ electron in a single-electron atom will be written as

$${}^2D_{5/2}, {}^2D_{3/2}.$$

(*b*) The level $^2D_{5/2}$ ($j = \frac{5}{2}$) corresponds to spin momentum parallel to orbital momentum while the level $^2D_{3/2}$ ($j = \frac{3}{2}$) corresponds to spin momentum anti-parallel to orbital momentum. Of these, the more stable level is one in which the magnetic moment of the electron spin, $\vec{\mu}_s$, lines up in the direction of the magnetic field $\vec{B}$ produced by the orbital motion of the electron.

Now, the field $\vec{B}$ due to orbital motion is always in the direction of angular momentum vector $\vec{L}$ (Fig.). Since the electron is negatively charged, the spin magnetic moment $\vec{\mu}_s$ is opposite to spin angular momentum $\vec{S}$. Thus $\vec{\mu}_s$ is in the direction of $\vec{B}$ for the level in which $\vec{S}$ is anti-parallel to $\vec{L}$ *i.e.* for the level corresponding to $j = l — s$ (Fig.). ($\vec{\mu}_s$ is opposite to $\vec{B}$ when $\vec{S}$ is parallel to $\vec{L}$ as in Fig.). Hence the level $^2D_{3/2}$ is more stable *i.e.* of lower energy. The level $^2D_{5/2}$ is higher.

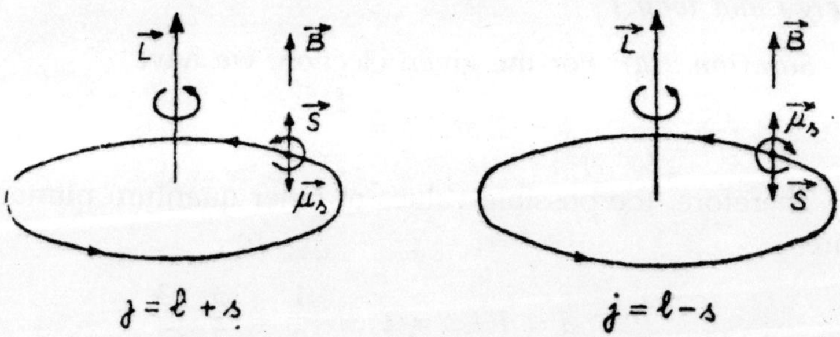

(*c*) When the atom is placed in a weak magnetic field, each energy level breaks up into $2j + 1$ magnetic levels corresponding to $mj = j, j—1, \ldots\ldots 0, \ldots\ldots — j$. Thus the level $^2D_{5/2}$ breaks up into 6 magnetic levels $\left(m_j = \frac{5}{2}, \frac{3}{2}, \frac{1}{2}, -\frac{1}{2}, -\frac{3}{2}, -\frac{5}{2}\right)$ and the level $^2D_{3/2}$ breaks up into 4 magnetic levels $\left(m_j = \frac{3}{2}, \frac{1}{2}, -\frac{1}{2}, -\frac{3}{2}\right)$.

(*d*) The level corresponding to $m_j = \frac{5}{2}$ involves highest magnetic shift and hence lies highest.

4. *For an electron orbit with quantum number l = 2, state the possible values of the components of angular momentum along a specified direction.*

Solution : The component of total angular momentum along a specified direction (z-axis) is quantised, and takes values given by

$$J_z = m_j \cdot \frac{h}{2\pi}$$

where m_j is the magnetic quantum number corresponding to the inner quantum number j.

For the given electron, we have

$$l = 2 \text{ (d-electron)}$$

and, of course $s = \frac{1}{2}$.

The two possible values of j are

$$j = l \pm s$$

$$= 2 \pm \frac{1}{2} = 5/2 \text{ and } 3/2.$$

For $j = 5/2$ the possible values of mj are

$$mj = 5/2,\ 3/2,\ 1/2,\ -1/2,\ -3/2,\ -5/2.$$

For $j = 3/2$ the possible values of m_j are

$$m_j = 3/2,\ 1/2,\ -1/2,\ -3/2.$$

The values differ by integers. Thus, the possible values of the z-components of total angular momentum are

$$\pm\frac{5}{2}\left(\frac{h}{2\pi}\right), \pm\frac{3}{2}\left(\frac{h}{2\pi}\right), \pm\frac{1}{2}\left(\frac{h}{2\pi}\right).$$

5. *Calculate the possible orientations of the total angular momentum vector $\vec{J}$ corresponding to $j = 3/2$ with respect to a magnetic field along the z-axis.*

Solution : The magnitude of the total angular momentum and its z-component are quantised according to the relations

$$J = \sqrt{[j(j+1)]}\,\frac{h}{2\pi}$$

and $$J_z = m_j \frac{h}{2\pi}$$

For $j = 3/2$, we have

$m_j = 3/2, 1/2, -1/2, -3/2.$

The angle θ between $\vec{J}$ and the z-axis is determined by m_j and j, according as

$$\cos\theta = \frac{J_z}{J} = \frac{m_j}{\sqrt{[j(j+1)]}}$$

Now $$\sqrt{[j(j+1)]} = \sqrt{\left[\frac{3}{2}\left(\frac{3}{2}+1\right)\right]} = \frac{\sqrt{(15)}}{2}.$$

$$\therefore \quad \cos\theta = \frac{2m_j}{\sqrt{(15)}}$$

For $mj = 3/2, 1/2, -1/2, -3/2$, we have

$\cos\theta = 0.775, 0.258, -0.258, -0.775$

$\therefore \quad \theta = 39.2°, 75.0°,]114°, 140.8°.$

The orientations of $\vec{J}$ with respect to z-axis are shown in Fig.

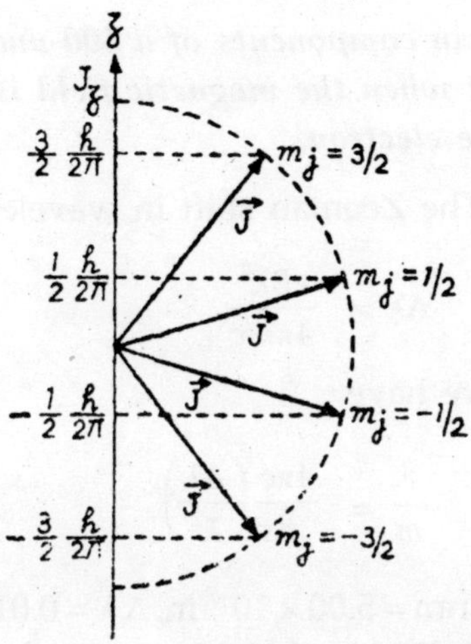

6. *An element under spectroscopic examination is placed in a magnetic field of flux density 0.3 weber/meter^2. Calculate the Zeeman shift of a spectral line of wavelength 4500 Å.*

Solution : The Zeeman shift in frequency is given by

$$\Delta v = \frac{eB}{4\pi m}$$

where e and m are the charge and mass of electron. Since $v = \frac{c}{\lambda}$, we have

$$\Delta v = -\frac{c}{\lambda^2}\Delta\lambda$$

$$\therefore \qquad \Delta\lambda = \frac{\lambda^2}{c}\Delta v = \frac{eB\lambda^2}{4\pi mc}.$$

Here $B = 0.3$ weber/meter^2, $\lambda = 4500$ Å $= 45 \times 10^{-7}$ meter.

Thus

$$\Delta\lambda = \frac{(1.6 \times 10^{-19} \text{ coul}) \times (0.3 \text{ weber/meter}^2) \times (4.5 \times 10^{-7} \text{ m})^2}{4 \times 3.14 \times (9.1 \times 10^{-31} \text{kg})(3 \times 10^8 \text{ meter/sec})}$$

$$= 0.0283 \times 10^{-10} \text{ meter} = 0.0283 \text{ Å}.$$

7. The Zeeman components of a 500-nm spectral line are 0.0116 nm apart when the magnetic field is 1.00 T. Find the ratio e/m for the electron.

Solution : The Zeeman shift in wavelength is given by

$$\Delta\lambda = \frac{eB\lambda^2}{4\pi mc}.$$

From this, we have

$$\frac{e}{m} = \frac{4\pi c}{B}\left(\frac{\Delta\lambda}{\lambda^2}\right).$$

Here $\lambda = 500$ nm $= 5.00 \times 10^{-7}$ m, $\Delta\lambda = 0.0116$ nm $= 0.000116 \times 10^{-7}$ m, B =1.00, Tesla = 1.00 N/A-m and also c = 3.0×10^8 m/s.

$$\therefore \quad \frac{e}{m} = \frac{4 \times 3.14\left(3.0 \times 10^8 \text{ m/s}\right)\left(0.0001 \times 10^{-7} \text{ m}\right)}{\left(1.00 \text{ N/A}-\text{m}\right)\left(5.00 \times 10^{-7} \text{ m}\right)^2}$$

$$= 1.75 \times 10^{11} \text{ C/kg}.$$

12

Hydrogen Spectrum

The spectrum of hydrogen atom consists of a number of lines. These lines have been grouped into a number of 'series'. The lines in each series are such that *their separation and intensity decrease regularly towards shorter wavelengths,* converging to a limit called the 'series limit'.

The wavelengths in each series can be given by a simple empirical formula. The first such spectral series was observed by Balmer in 1885 and is called the Balmer series of hydrogen. The first line with the longest wavelength (6563 A) is named $H\alpha$, the next $H\beta$, and so on. The series limit lies at 3646 A, beyond which is a faint continuous spectrum. Balmer's formula for the wavelengths of the series is

$$\frac{1}{\lambda} = R\left(\frac{1}{2^2}-\frac{1}{n^2}\right), n = 3, 4, 5, \ldots \ldots \quad \text{(Balmer)}$$

The quantity R is called the 'Rydberg constant' and has the value

$$R = 1.097 \times 10^7 \text{ meter}^{-1}.$$

The H_α line corresponds to $n = 3$, the H_β line to $n = 4$, and so on. The series limit corresponds to $n = \infty$, so that it occurs at a wavelength of $4/R$, in agreement with experiment.

The Balmer series contains only those spectral lines which fall in the visible part of the hydrogen spectrum. The lines falling in the ultraviolet and infrared parts form other series. The lines in the ultraviolet form the Lyman series whose wavelengths are given by

$$\frac{1}{\lambda} = R\left(\frac{1}{1^2} - \frac{1}{n^2}\right), n = 2, 3, 5, \dots \dots \qquad \text{(Lyman)}$$

In the infrared, three spectral series have been observed whose lines have the wavelengths given by the formulas

$$\frac{1}{\lambda} = R\left(\frac{1}{3^2} - \frac{1}{n^2}\right), n = 4, 5, 6, \dots \dots \qquad \text{(Paschen)}$$

$$\frac{1}{\lambda} = R\left(\frac{1}{4^2} - \frac{1}{n^2}\right), n = 5, 6, 7, \dots \dots \qquad \text{(Brackett)}$$

$$\frac{1}{\lambda} = R\left(\frac{1}{5^2} - \frac{1}{n^2}\right), n = 6, 7, 8, \dots \dots \qquad \text{(Pfund)}$$

Bohr's Theory of Hydrogen Spectrum : The existence of sharp spectral lines cannot be explained from the classical electromagnetic theory. Bohr explained it by applying Planck's quantum hypothesis to the Rutherford's atomic model. The Rutherford's atom consists of a central massive nucleus containing the positive charge of the atom, and the electrons move round the nucleus in circular planetary orbits. The centripetal force required for the orbital motion is provided by electrostatic attraction between the positively-charged nucleus and the negatively-charged electron.

Bohr Proposed Two Postulates

(i) An electron can move *only* in those orbits for which the angular momentum L of the electron is an integral multiple of $h\backslash 2\pi$ where h is Planck's constant. (Thus Bohr quantised the angular momentum of the electron.) The electron moving in any of the permitted orbits does not radiate energy in spite of its acceleration towards the centre of the orbit. The atom, therefore, is said to exist *in* a *stationary* state.

(ii) The emission (or absorption) of radiation by the atom lakes place when an electron jumps from one permitted orbit to another. The radiation is emitted (or absorbed) as a single quantum (photon) whose energy hv is equal to the difference in energies of the electron in the two orbits involved. Thus if Ei be the energy of the initial orbit of the electron and Ef that of the final orbit, then we have

$$hv = E_i - E_f$$

where v is the frequency of the emitted (or absorbed) radiation.

Let e, m and v be the charge, mass and velocity of the electron (measured in coulomb, kg and meter/sec respectively) and r the radius of the orbit measured in meter. The positive charge on the nucleus is Ze, where Z is the atomic number (Fig.). In case of hydrogen Z=l, so that positive charge on nucleus is e. As the centripetal force is provided by the electrostatic attraction, we have

$$\frac{mv^2}{r} = \frac{1}{4\pi_0^{\epsilon}} \frac{e^2}{r^2}$$

or $$mv^2 = \frac{e^2}{4\pi_0^{\epsilon} r}. \qquad \ldots (i)$$

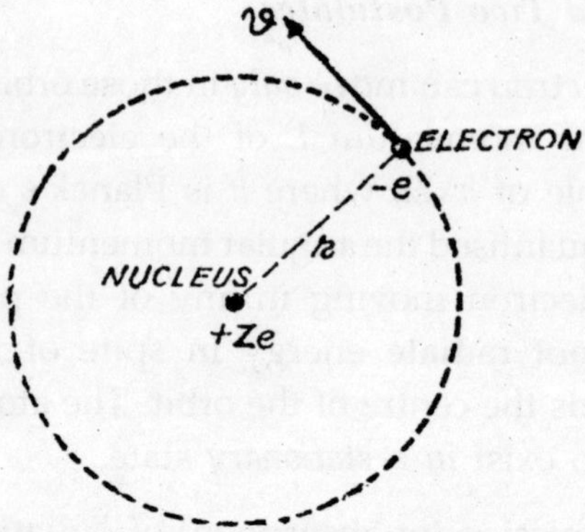

From the first postulate, the angular momentum of the electron is given by

$$L = mvr = n\,\frac{h}{2\pi},$$

where n called as 'quantum number' = 1, 2, 3,...

Squaring eq. (*ii*) and dividing by eq. (*i*), we get

$$\boxed{r = n^2\,\frac{h^2 \epsilon_o}{\pi m e^2},}\quad n = 1, 2, 3, \ldots \qquad \ldots (iii)$$

This is the expression for the radius of the permitted orbits.

Now, the energy E of the electron in an orbit is the sum of kinetic and potential energies. The kinetic energy of the electron

$$K = \frac{1}{2}\,mv^2 = \frac{e^2}{8\pi\epsilon_0 r}. \qquad \text{[form eq. } (i)]$$

The potential energy at a distance r from the nucleuses equal to the work done in removing the electron from r to infinity *against the electrostatic attraction* $\left(-\frac{e^2}{4\pi\epsilon_0 r^2}\right)$, and is given by

$$U = \int_r^\infty -\frac{e^2}{4\pi\epsilon_0 r^2}\,\mathrm{dr} = \frac{1}{4\pi\epsilon_0}\left[\frac{e^2}{r}\right]_r^\infty = -\frac{e^2}{4\pi\epsilon_0 r}.$$

Hence the total energy of the electron is

$$E = K + U = \frac{e^2}{8\pi\epsilon_0 r} - \frac{e^2}{4\pi\epsilon_0 r} = -\frac{e^2}{8\pi\epsilon_0 r}.$$

Substituting for r from eq. (iii), we get

$$\boxed{E = -\frac{\mathrm{me}^4}{8\epsilon_0^2 h^2}\left(\frac{1}{n^2}\right);} \quad n = 1, 2, 3, \ldots \qquad \ldots(iv)$$

This is the expression for the energy of the electron in the nth orbit. We see that it is negative.

Let E_i and E_f be the energies of the election corresponding to the initial (higher) and final (lower) orbits of the excited atom. Then we have

$$E_i = -\frac{me^2}{8\epsilon_0^2 h^2}\left(\frac{1}{n_i^2}\right);$$

and
$$Ef = -\frac{me^4}{8\epsilon_0^2 h^2}\left(\frac{1}{n_f{}^2}\right)$$

where n_i and n_f are the corresponding quantum numbers. The energy difference between these states is

$$E_i - E_f = \frac{me^4}{8\epsilon_0^2 h^2}\left(\frac{1}{n_f^2} - \frac{1}{n_i^2}\right).$$

Hence, from Bohr's second postulate, the frequency v of the emitted photon is

$$v = \frac{E_i - E_f}{h}$$

$$= \frac{me^4}{8\epsilon_0^2 h^3}\left(\frac{1}{n_f^2} - \frac{1}{n_i^2}\right).$$

The corresponding wavelength A is given by

$$\frac{1}{\lambda} = \frac{v}{c} = \frac{me^4}{8\epsilon_0^2 ch^3}\left(\frac{1}{n_f^2} - \frac{1}{n_i^2}\right)$$

This equation indicates that, *since n_i and nf can take only integral values, the radiation emitted by excited hydrogen atoms should contain certain discrete wavelengths only.*

The value of the constant *term* $\frac{me^4}{8\epsilon_0^2 ch^3}$ comes out to be the same as the Rydberg constant R in the Balmer's empirical formula. Thus we have

$$\frac{1}{\lambda} = R\left(\frac{1}{n_f^2} - \frac{1}{n_i^2}\right).$$

Emission of Spectrum

When the hydrogen atom gets sufficient energy from outside by some means, the electron from an inner orbit of lower energy goes up to an outer orbit of higher energy. This excited state of the atom lasts for about 10 ^{8} second after which the electron jumps back to an inner orbit. In jumping down, it emits the difference in energy between two orbits as electromagnetic radiation. If the electron jumps from an orbit n_i to an orbit n_f, the wavelength of the emitted radiation will be

$$\frac{1}{\lambda} = R \cdot \left(\frac{1}{n_f^2} - \frac{1}{n_i^2}\right).$$

It is found that for

$n_f = 1,\ n_i = 2, 3, 4, \ldots$ we obtain Lyman series,

$n_f = 2, n_i = 3, 4, 5, \ldots$ we obtain Balmer series,

$n_f = 3, n_i = 4, 5, 6, ...$, we obtain Paschen series,

$n_f = 4, n_i = 5, 6, 7, ...$ we obtain Brackett series,

$n_f = 5, n_i = 6, 7, 8, ...$ we obtain Pfund series.

The corresponding energy level diagram is shown in Fig. below. The top horizontal line represents zero energy *i.e.* energy of the electron outside the atom ($n = \infty$). The other horizontal lines represent energies of different orbits given by the formula

$$E = -\frac{me^4}{8\epsilon_0^2 h^2}\left(\frac{1}{n^2}\right).$$

The arrows ending at the lines $n = 1, 2, 3, 4$ and 5 represent the transitions responsible for the Lyman, Balmer, Paschen, Brackett and Pfund series respectively.

Shortcomings of Bohr's Theory : Bohr's theory although very successful in explaining the hydrogen spectrum and giving valuable information about atomic structure, has the following shortcomings :

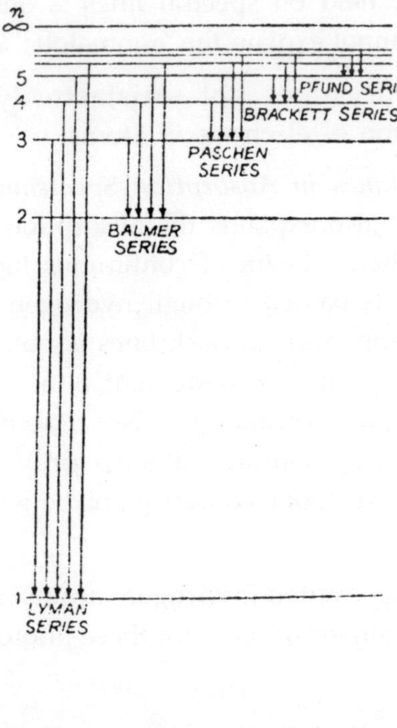

1. An individual line of hydrogen spectrum, when examined under a high-resolving spectroscope is found to be accompanied by a number of faint lines. This 'fine structure' of spectral lines cannot be explained by Bohr's theory as such. It can, however, be explained when
   (i) the relativistic variation in the mass of the electron and
   (ii) the electron 'spin' are taken into account.
2. Bohr's theory cannot explain the variation in intensity of the spectral lines of an element. The intensity can be explained by quantum mechanics.
3. The theory is only applicable to one-electron atoms such as hydrogen, hydrogen isotopes, singly-ionized helium, doubly-ionised lithium, etc. It does not explain the spectra of complex atoms.
4. The success of Bohr's theory in explaining the effect of magnetic field on spectral lines is only partial. The theory cannot explain the 'anomalous' Zeeman effect'.
5. The theory does not satisfactorily explain the distribution of electrons in atoms.

No Balmer Lines in Absorption Spectrum of Hydrogen: The Bohr theory also explains the absorption line spectrum of hydrogen. When a beam of continuous light (containing all wavelengths) is passed through hydrogen and then sent into a spectrograph, a set of dark lines is obtained. In terms of quantum theory, the incident light is a beam of quanta (photons) of all sorts of energies. Now, according to Bohr theory, the hydrogen atoms absorb only those quanta whose energies correspond to transitions between its discrete energy levels.

The resulting excited hydrogen atoms re-radiate the absorbed energy almost at once but these photons come off in random directions with only a few in the same direction as the original beam of continuous light. The dark lines in the absorption spectrum are therefore never completely black. Obviously the absorption lines will have exactly the same frequencies as the emission lines.

Now it is found that all the emission lines of hydrogen spectrum do not appear in the absorption spectrum. The reason is that normally the atom is always in the ground state $n = 1$. Therefore, absorption transitions can only occur from $n = 1$ to n>l. Hence lines of only the Lyman series can appear in absorption spectrum. To obtain Balmer series in absorption, the atom must initially be in the state $n = 2$, because Balmer lines require transitions from $n = 2$ to $n = 2$. Since atoms are usually in the ground state, Balmer lines are not obtained in absorption.

Assumptions of Planck, Einstein and Bohr: Planck had assumed that the atoms of a hot body behave as oscillators and have discrete (quantised) energies. They do not emit radiant energy continuously, but only in 'jumps' or 'quanta'. Planck, however, still maintained that radiation propagates continuously through space as electromagnetic waves.

Einstein, in order to explain the photoelectric effect, went a step ahead. He proposed that the radiation not only is emitted as quantum at a time, but also propagates as individual quanta (photons). He thus treated the propagation of radiation as particle propagation rather than wave propagation.

Bohr, in order to explain the hydrogen spectrum, adopted the Planck's quantum hypothesis that the radiation is emitted discontinuously from the atom. He started with the quantisation of the angular momentum of the electron in the orbit which ultimately results in the quantisation of energy of the atom.

Negative Energy of Hydrogen Orbits

An electron revolving in a hydrogen orbit has a negative potential energy by virtue of its attraction towards the nucleus, and also kinetic energy (which is positive) by virtue of its motion. The potential energy is greater in magnitude than the kinetic energy, so that the net energy is negative. The 'negative' energy signifies that the electron cannot escape from the atom. A positive energy for a nucleus-electron combination would mean that electron is not bound to the nucleus. Such a combination cannot constitute an atom.

Absorption of a Photon of Energy Greater than the Binding Energy : The binding energy of the hydrogen atom is the energy which must be supplied to it in order to remove its electron from the lowest state (level) to a state of zero total energy. It is numerically equal to the energy of the lowest state (: 136 eV).

The highest energy level of the atom corresponds to n=∞ so that £=0. Above this are the energy states of the system consisting of an unbound electron plus the ionized atom. The total energy of an unbound electron is not quantised and is positive. Thus any energy $E > 0$ is possible for the electron, and the energy states form a continuum. When the hydrogen atom receives a photon of *any* energy which is greater than its binding energy of 13.6 eV, it absorbs the photon and the electron of the atom passes from its discrete ground state to a continuum state.

Bohr Theory as Corrected for Nuclear Mass: In Bohr theory we assume that the nucleus of the hydrogen atom is infinitely heavy compared with electron, and so it remains stationary while the electron revolves around it. In fact, the nucleus has a *finite* mass, and both the electron and the nucleus revolve about their common centre of mass C (Fig.) with a common angular velocity ω. Let us make correction in Bohr theory for the finite mass (motion) of the nucleus.

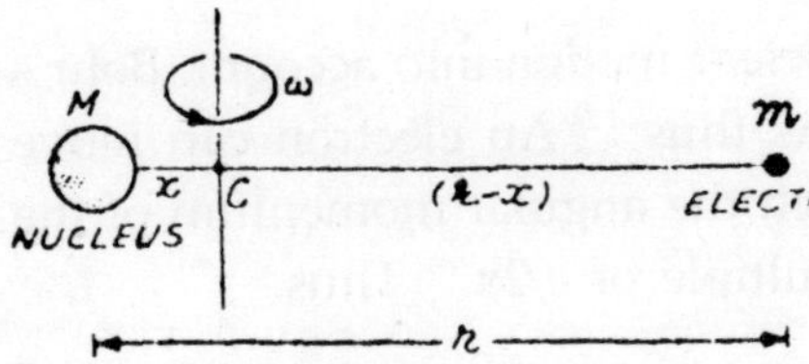

Let m be the mass of electron, M_H that of hydrogen nucleus, and r the distance between them. Let x be the distance of the nucleus from the common centre of mass C. The distance of the electron will be $(r—x)$. Since the system is in equilibrium, the moments of M_H and m about C will be equal. That is

$$M_H\, x = m\,(r - x).$$

Solving we get: $x = \dfrac{mr}{M_H + m}$

and $r - x = \dfrac{M_H\, r}{M_H + m}$

The total angular momentum of the atom about the centre of mass is

$$L = M_H\, x^2\, \omega + m\,(r - x)^2\, \omega$$

$$= M_H \frac{m^2 r^2 \omega}{(M_H + m)^2} + m \frac{M_H^2\, r^2 \omega}{(M_H + m)^2}$$

$$= M_H \frac{m r^2 \omega}{(M_H + m)^2} (m + M_H)$$

$$= \left(\frac{M_H\, m}{M_H + m}\right) r^2 \omega$$

$$= \mu\, r^2\, \omega,$$

where $\mu \left(\dfrac{M_H\, m}{M_H + m}\right)$ is called the 'reduced mass' of the electron, because its value is slightly less than m.

Taking nuclear motion into account, Bohr's first postulate will be read as thus : "An electron can move only in those orbits for which the angular momentum of the whole *atom* is an integral multiple of $h/2\pi$" Thus

$$\mu r^2\text{–}\omega = n\,\frac{h}{2\pi}. \qquad \text{... } (i)$$

In the absence of nuclear motion, the corresponding equation was

$$mvr = n\,\frac{h}{2\pi}.$$

But $v = r\omega$, so that

$$mr^2\omega = n\,\frac{h}{2\pi}. \qquad \text{...}(ii)$$

Eq. (*i*) is exactly the same as eq. (*ii*) except that μ has replaced *m*. Therefore all the results of Bohr theory (which assume mass of the nucleus infinite) still hold, provided *m* is replaced by μ.

The energy of the electron in the *n*th orbit is now

$$E = -\frac{\mu e^4}{8_0^{\epsilon^2} h^2}\left(\frac{1}{n^2}\right).$$

Since μ is slightly less than *m*, the electron energies are slightly less negative than if the nucleus were at rest (*i.e.* infinitely heavy). The wavelengths of the hydrogen lines are now given by

$$\frac{1}{\lambda} = \frac{\mu e^2}{8_0^{\epsilon^2} ch^3}\left(\frac{1}{n_{f^2}} - \frac{1}{n_i^2}\right)$$

$$= R_H\left(\frac{1}{n_f^{\ 2}} - \frac{1}{n_i^2}\right),$$

where $R_H \left(\dfrac{\mu e^4}{8\epsilon_0^2 ch^3}\right)$ is the Rydberg constant *for the hydrogen atom.* Since the value of the Rydberg constant is slightly less than if the nucleus were at rest, *the wavelengths are slightly larger than those corresponding to an infinitely heavy nucleus.*

Effect of Finite Nuclear Mass on Rydberg Constant: Taking nuclear motion into account, the Rydberg constant *for the hydrogen atom* is

$$R_H = \frac{\mu e^4}{8\epsilon_0^2 ch^3}$$

Now
$$\mu = \frac{M_H m}{M_H + m}.$$

$$\therefore \quad R_H = \frac{me^4}{8\epsilon_0^2 ch^3}\left(\frac{M_H}{M_H + m}\right).$$

But $\dfrac{me^4}{8\epsilon_0^2 ch^3}$ is the value of the Rydberg constant of an infinitely heavy nucleus, and is a constant *for all the atoms.* Let it be denoted by R_∞. Its value is 109737 cm^{-1}. Thus

$$R_H = R\infty\left(\frac{M_H}{M_H + m}\right) = \frac{R\infty}{1 + (m / M_H)}.$$

This shows that the Rydberg constant for hydrogen, R_H depends upon the mass M_H of the hydrogen nucleus. Thus it is a constant only for a particular atom, and varies slightly from atom to atom. This variation, clearly, arises because of the finite mass (motion) of the nucleus. The Rydberg constant for hydrogen is $R_H = 109677$ cm^{-1}.

The variation of Rydberg constant from element to element is shown in figure below. The largest change occurs between H^1 and H^2. With increasing mass number, the Rydberg constant approaches more and more closely to R_∞.

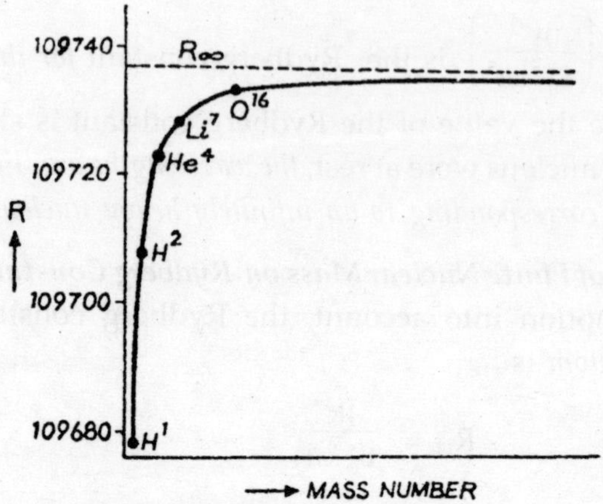

Discovery of Heavy Hydrogen

The variation of Rydberg constant with the mass of the nucleus resulted in the discovery of deuterium (heavy hydrogen), The deuterium is an isotope of hydrogen whose nucleus has a mass almost exactly double that of ordinary hydrogen. Therefore, the Rydberg constant for deuterium is slightly greater than that for hydrogen. Consequently, the wavelengths of the spectral lines of deuterium are slightly shorter than those of the corresponding spectral lines of hydrogen. The general nature of the two spectra is, however, exactly same because atoms of both hydrogen and deuterium have the same electron structure.

Urey, in 1932, photographed hydrogen spectrum by means of a large concave grating. He observed that each hydrogen line was accompanied by a very faint line slightly on the short wavelength side. He measured the wavelength difference and attributed the faint line to some isotope of hydrogen. This isotope is now known as 'heavy hydrogen' or 'deuterium.

Bohr Theory Applied to Singly-ionised Helium Atom (He^+): The Bohr theory can be applied to ions containing single electron such as He^+ (singly-ionised helium), Li^{++} (doubly-ionised lithium), and so on. Such ions behave just like hydrogen atom except that they have greater nuclear charge and mass. Hence they are also called as 'hydrogen-like atoms'.

The nucleus of a hydrogen-like atom of atomic number Z carries a charge +*Ze* so that the electrostatic force of attraction which it exerts on the orbital electron is $\frac{1}{4\pi\epsilon_0}\frac{Ze^2}{r^2}$, instead of $\frac{1}{4\pi\epsilon_0}\frac{e^2}{r^2}$ as in the case of hydrogen atom for which Z = 1. Consequently, the formula for the energy levels of a hydrogen-like atom is

$$E = -\frac{\mu Z^2 e^4}{8\epsilon_0^2 h^2}\left(\frac{1}{n^2}\right),$$

where μ is the reduced mass of the electron in the hydrogen-like atom, which is different than that in *H*. The corresponding wave-length JA are given by

$$\frac{1}{\lambda} = \frac{\mu Z^2 e^4}{8\epsilon_0^2 ch^3}\left(\frac{1}{n_f^2}-\frac{1}{n_i^2}\right),$$

Now for He^+, we have Z = 2 and $\frac{\mu e^4}{8\epsilon_0^2 ch^3} = R_H$, (the Rydberg constant for ionised helium), μ now being the reduced mass of the electron in He^+. Therefore

$$\frac{1}{\lambda} = 4R_{He}\left(\frac{1}{n_f^2}-\frac{1}{n_i^2}\right),$$

The formula for hydrogen (Z = 1) is

$$\frac{1}{\lambda} = R_H\left(\frac{1}{n_f^2}-\frac{1}{n_i^2}\right),$$

Relationship between H and He⁺ Spectra : The formula for He^+ differs from the formula for hydrogen by a factor of 4, provided we ignore the difference between R_{He} and R_H. The corresponding helium wavelengths are therefore shorter by a factor of 4 . It is clear from the above two formulae that a transition from $n_i = 4$ to $n_f = 2$ in He$^+$ would emit a photon of the same wavelength as that emitted during a transition from n_i, = 2 to $n_f = 1$ in H. Hence we can conclude that the He$^+$ spectral lines coincide with certain H spectral lines. The coincident is, however, not exact because the Rydberg constant, for H and He$^+$ are slightly different.

Below given figure shows Balmer series of H and the Pickering series of He$^+$ photographed on the same plate. We see that alternate members of Pickering series of He$^+$ coincide roughly with members of Balmer series of H.

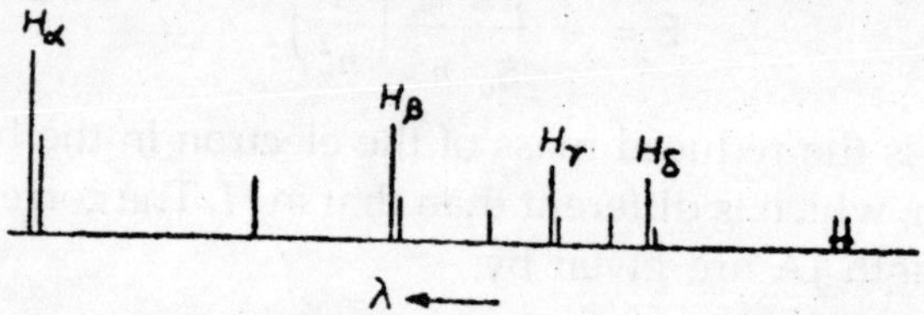

Determination of m/M$_H$:— The Rydberg constant for hydrogen atom is given by

$$R_H = \frac{R\infty}{1+(m/M_H)}, \qquad \text{... } (i)$$

where R_∞ is the Rydberg constant for an infinitely heavy nucleus, m is the mass of electron and M_H is the mass of hydrogen nucleus (*i.e.* mass of proton). Similarly, the Rydberg constant for He$^+$ is given by

$$R_{He} = \frac{R\infty}{1+(m/M_{He})},$$

where M_H, is the mass of a helium nucleus. Since helium nucleus is very nearly equal to four times hydrogen nucleus ($M_H \simeq 4M_H$), we write

$$R_{He} = \frac{R\infty}{1+(m/4M_H)}. \qquad \text{... } (ii)$$

Dividing eq. (*i*) by (*ii*), we get

$$\frac{R_H}{R_{He}} = \frac{1+(m/4M_H)}{1+(m/M_H)}$$

or $$R_H\left(1+\frac{m}{M_H}\right) = R_{He}\left(1+\frac{m}{4M_H}\right)$$

or $$\frac{m}{m_H}\left(R_H-\frac{1}{4}R_{He}\right) = R_{He}-R_H$$

or $$\frac{m}{M_{H}} = \frac{R_{He}-R_H}{R_H-\frac{R_{He}}{4}}.$$

R_{He} and R_H can be found by measurements of wavelengths in the series of helium and hydrogen spectra. Hence m/M_H, the ratio of the mass of electron to the mass of proton (*i.e.* hydrogen nucleus) can be determined.

Sommerfeld's Extension of Bohr Theory: According to Bohr's theory, an atom has a number of discrete energy levels (orbits), which only can be occupied by electrons. Each level is characterised by an integer *n* called the 'quantum number.' In a hydrogen atom, the (quantised) energy of a level with quantum number *n* is given by (ignoring nuclear motion)

$$En = -\frac{m e^4}{8\epsilon_0^2 h^2}\left(\frac{1}{n^2}\right),$$

where *m*, *e* and *h* have their usual meanings. This we can write as

$$En = -\frac{R hc}{n^2} \quad \text{... } (i)$$

where $R\left(=\frac{me^4}{8\epsilon_0^2 ch^3}\right)$ is the Rydberg constant and *c* is the velocity of light. Whenever the electron jumps from one level to another, electromagnetic radiation of a *discrete* frequency is emitted (or absorbed). For example, when the electron jumps from the $n = 3$ to $n = 2$ level, the first line of Balmer series (called H_α-line) is emitted (see Fig.) and its frequency is

$$v = \frac{E_3 - E_2}{h}.$$

in a similar way, the emission of all spectral lines can be explained.

According to the above theory, each line of the hydrogen spectrum must be a *single* line *i.e.* it must have a single frequency (or wavelength). It was, however, observed by Michelson and others that in fact, the hydrogen lines were not strictly single. Each line consisted of a small number of close components of slightly different frequencies. This is called the 'fine structure' of spectral lines. This infers that each energy level of a given n must be splitted into sub-levels of slightly different energies. Hence the fine-structure could not be explained by Bohr's original theory,

Sommerfeld extended the Bohr's theory by assuming the existence of elliptical orbits (in addition to Bohr's circular orbits) for the electron, with one focus at the nucleus of the atom (Fig.). Now, an electron in an elliptic orbit has *two* degrees of freedom, namely the radial distance r and the azimuthal angle θ. Sommerfeld postulated that *each of these degrees of freedom must be quantised separately*. If $p\theta$ and p, be the angular and radial momenta of the electron, then according to Wilson Sommerfeld quantisation rules ; we must have

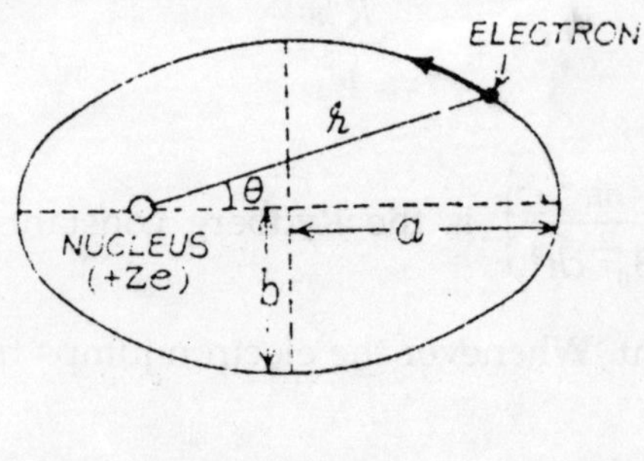

$$\int_0^{2x} pe\, d\theta = \text{kh}, \qquad \text{k} = 1, 2, 3, \ldots \ldots \qquad \ldots (ii)$$

and $$\oint p_r\, dr = n_r\, h, \qquad \text{nr} = 0, 1, 2, \ldots \ldots \qquad \ldots (iii)$$

where k and n_r are integers, called respectively 'azimuthal' and 'radial' quantum numbers. By integrating over a complete revolution, it can be proved that

$$1 - e^2 = k^2 / K + n_r)^2$$

where e is the eccentricity of the ellipse. Since both k and n, ate integers, we can put

$$k + n_r = n, \qquad (n = 1, 2, 3, \ldots\ldots)^*$$

where n is called the 'total quantum number' of the electron. Thus

$$1 - \epsilon^2 = \frac{k^2}{n^2}.$$

If a and b be the semi-major and semi-minor axes of the ellipse, then $(1 - e^2) = b^2\ a^2$. Therefore

$$\frac{b^2}{a^2} = \frac{k^2}{n^2}$$

or $$\boxed{\frac{b}{a} = \frac{k}{n}.} \qquad \ldots (iv)$$

This is the condition of quantisation for the orbits. *Only those elliptic orbits are permitted for the electron for which the ratio of the major to the minor axis is the ratio of two integers.*

Let us now calculate the total energy E of an electron in a quantised elliptic orbit. The energy is the sum of the kinetic energy K and potential energy U, That is

$$E = K + U$$

$$= \frac{1}{2} m \,(r^2 + r^2\theta^2) - \frac{1}{4\pi_0^{\epsilon}} \frac{e^2}{r}.$$

But $p_r = mr$ and $p_\theta = mr^2\theta$. Therefore the last expression may be written as

$$E = \frac{1}{2m}\left\{p_r^2 + \frac{p_\theta^2}{r^2}\right\} - \frac{1}{4\pi_0^{\epsilon}} \frac{e^2}{r}.$$

$$\text{Now, } p_r = mr = m\,\frac{dr}{dt} = m\,\frac{dr}{d\theta}\frac{d\theta}{dt} = m\,\frac{dr}{d\theta}\,\theta = (mr^2\theta)\,\frac{1}{r^2}\frac{dr}{d\theta}$$

$$= \frac{p_\theta}{\gamma^2}\frac{dr}{d\theta}$$

Putting this value of p_r in the last eq., we get

$$E = \frac{p_\theta^2}{2mr^2}\left[\left(\frac{1}{r}\frac{dr}{d\theta}\right)^2 + 1\right] - \frac{1}{4\pi_0^{\epsilon}}\frac{e^2}{r}. \quad \text{... } (v)$$

Now, the polar equation of the ellipse is

$$\frac{1}{r} = \frac{1}{a}\frac{1 - \epsilon\cos\theta}{1 - \epsilon^2} \quad \text{... } (vi)$$

where $(1 - \epsilon^2) = b^2/a^2$. If we put the value of r and $\dfrac{dr}{d\theta}$ from eq. (*vi*) in eq. (*v*), then we can show that

$$E = -\frac{me^4\left(1 - \epsilon^2\right)}{\left(4\pi_0^{\epsilon}\right)^2 2p_\theta^2}. \quad \text{... } (vii)$$

For an isolated system, the angular momentum p_θ is constant. Then from eq. (*ii*), we get

$$p_\theta = \frac{kh}{2\pi}.$$

$$\text{Also} \quad (1 - \epsilon^2) = \frac{b^2}{a^2} = \frac{k^2}{n^2}. \quad \text{[from eq. } (iv)]$$

Making these two substitutes in eq, (vii), we get

$$E = -\frac{me^4\left(\frac{k^2}{n^2}\right)}{\left(4\pi\epsilon_0\right)^2 2\left(\frac{kh}{2\pi}\right)^2}$$

$$= -\frac{me4}{8\epsilon_0^2 h^2}\left(\frac{1}{n^2}\right).$$

Making the substitution, $R = \frac{me4}{8\epsilon_0^2 ch^3}$, we get

$$E = -\frac{Rhc}{n^2}. \qquad \text{... }(viii)$$

This is exactly the same as eq. (*i*). the energy of electron in a circular Bohr orbit.

Now, for a given value of n $(= k + n_r)$, k can take n possible different values (k = 1, 2, 3,.....n). This means that for a given n, there are n orbits of *different* eccentricities (according to the condition $\left(\frac{b}{a} = \frac{k}{n}\right)$ which can be occupied by the electron, But *the* energy E for all these orbits is the same, because E is independent of k and depends upon n only [see eq. (v*iii*)]. Thus *the mere introduction of elliptic orbits adds no new energy levels.*

Relativistic Correction

In an elliptic orbit the velocity of the electron varies, becoming very large when the electron passes close *to* the nucleus. The theory of relativity shows that when the velocity of a particle increases, its mass also increases. This effect makes the energy of the electron in a more elliptic (*i.e.* of greater eccentricity) orbit greater than that in a less elliptic orbit. Thus, *the different k-orbits with a given n have slightly different energies,* according to the following expression, also proved by Sommerfeld

$$E = \frac{Rhc}{n^2} - \frac{Rhc\,\alpha^2}{n^3}\left(\frac{1}{k} - \frac{3}{4n}\right),$$

where α is a pure number ($\simeq 1/137$) called the 'fine-structure constant'.

We can now explain the fine structure of H_α line which is emitted when the electron jumps from the $n = 3$ to $n=2$ level. The $n = 3$ level consists of 3 sub-levels corresponding to $k = 1$, 2, 3, and the $n=2$ level consists of 2 sub-levels corresponding to $k=1,2$ (Fig.). Only those electron-jumps from the upper sub-levels to the lower sub-levels are allowed for which k changes by unity ($\Delta k = \pm 1$). Thus there are three possible jumps, showing that the H_α-*line* actually consists of three components.

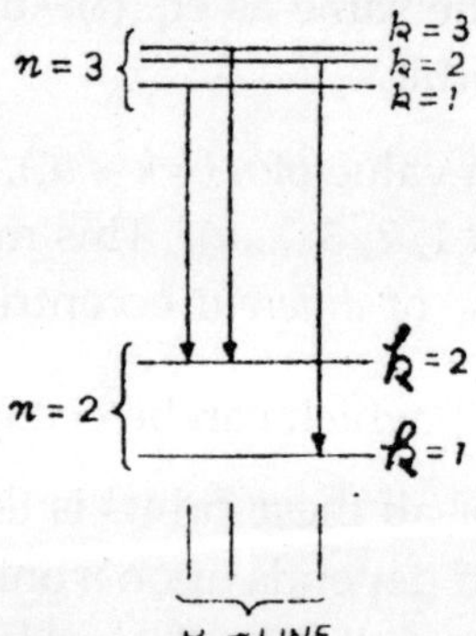

Note: From quantum mechanical considerations the azimuthal quantum number k has been replaced by $l + l$. Hence l can take the values 0, 1, 2, ... $(n - 1)$.

Excitation and Ionisation Potentials of an Atom: According to the Bohr's theory, in an atom there are certain discrete orbits only in which an electron can revolve without radiating energy. Each of these orbits of a given atom is characterised by a certain (quantised) energy. Hence we say that there are certain discrete energy levels in an atom.

When an electron in an atom absorbs sufficient energy from an outside source, it rises from its present energy level to a higher level. The atom is then said to be 'excited'. This excited state lasts only for about 10^{-8} seconds after which the electron jumps back to the inner level, emitting the absorbed energy in the form of electromagnetic radiation.

When the electron in the atom gains sufficient energy so as to escape *completely* from the atom, the atom is said to be 'ionised'.

Again when the atom recaptures an electron and reverts to its original state, the absorbed energy is emitted as electromagnetic radiation.

The energy required to excite or to ionise a given atom is perfectly definite. For example, in case of hydrogen atom the energy of the nth orbit is

$$E_n = -\frac{me^4}{8\epsilon_0 h^2}\left(\frac{1}{n^2}\right).$$

Substituting the known values of m ($=9.1\times10^{-31}$kg), e($=1.6\times10^{-19}$ coul), h($= 6.62\times10^{-34}$ joule-sec) and ϵ_0 ($= 8.85\times10^{-20}$ coul^2/nt-m^2), we get

$$E_n = -2.17\times10^{-18}\left(\frac{1}{n^2}\right) \text{ joule}$$

$$= -\frac{2.17\times10^{-18}}{1.6\times10^{-19}}\left(\frac{1}{n^2}\right)$$

$$= -\frac{13.6}{n^2} \text{ electron-volt.}$$

[1eV = 1.6×10^{-19} joule]

Putting $n = 1, 2, 3, ...\infty$, the energies of the 1st, 2nd, 3rd,......∞ energy levels come out to be —13.6, —3.4, —1.51,......0 electron-volts respectively. Hence the energy to be supplied to the atom to raise an electron from the first to the second orbit is

(13.6 – 3.4) = 10.2 electron-volts, from first to the third orbit is (13.6 – 1.51) = 12.09 electron-volts,* and to the ionized state is (13.6 – 0) = 13.6 electron-volts.

Now, the atoms are usually excited or ionised by bombarding them with electrons accelerated through a potential. But the excitation or unionisation of an atom can take place only when the bombarding electron has the required amount of energy.

The minimum accelerating potential which imparts to the bombarding electron an energy sufficient to excite a given atom is called the 'excitation potential' of the atom. Similarly, *the minimum accelerating potential which imparts to the bombarding electron an energy sufficient to ionise the atom is called the 'ionisation potential' of the atom.*

Thus 10.2, 12.09, electron-volts are the excitation potentials; while 13 6 electron-volts is the ionisation potential for the hydrogen atom. All these together are called the 'critical potentials' of the atom.

Franck-Hertz Experiment

Franck and Hertz, in 1914, performed a series of experiments to measure the excitation potentials of atoms of different elements. These experiments showed directly that in an atom discrete energy levels do exist.

Their apparatus is shown in Figure below. It consists of a glass tube in which are mounted a filament F, a grid G, and a plate P as shown. The filament F is heated by a small battery B. An accelerating potential V is applied between F and G ; and a *small fixed* retarding potential V_0 (about 0.5 volt) between G and P. The gas of the element whose atoms are to be studied is introduced in the tube at about 1 mm of mercury pressure.

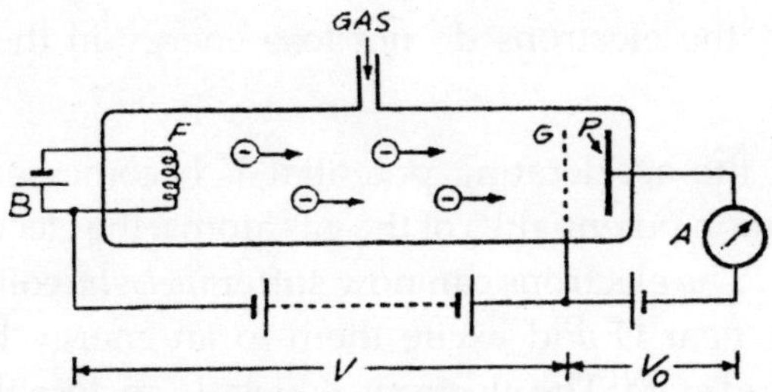

The electrons emitted from the hot filament *F* are accelerated between *F* and *G* by the potential *V*, and retarded between *G* and *P* by the potential V_0. Thus only electrons having energies greater than, eV_0 at *G* are able to reach *P*. The current to *P* is recorded by an ammeter *A*.

The current is plotted against the *accelerating* potential *V*. which is gradually increased from zero. The curve obtained shows a series of regularly-spaced peaks (Fig.).

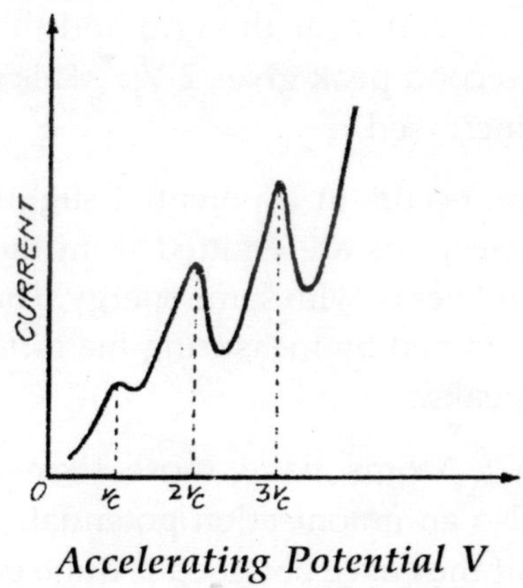

Accelerating Potential V

Interpretation of the Curve: Electrons are emitted from *F* with a range of small energies. In the beginning, they acquire a *small* additional energy e*V* on reaching *G*. Those electrons whose energy is now greater than eV_0 reach *P*, and a current is obtained. As the accelerating potential *V* is increased, more and more of the electrons arrive at the plate and the current rises. Between *F* and *G* the electrons collide with the gas atoms. But since the electron energies are not enough to excite the atoms, the electrons do not lose energy in these (elastic) collisions.

When the accelerating potential V becomes equal to the first excitation potential V_e of the gas atoms, the electron energy at G is cV_e. The electrons can now suffer *inelastic* collisions with gas atoms near G and excite them to an energy level above their ground state. The electrons which do so, lose their energy and are unable to reach P against the retarding potential. Hence the current drops sharply. Thus the position of the first peak gives the excitation potential V_c of the gas atoms.

As the accelerating potential V is increased further, the electrons suffer inelastic collisions nearer and nearer to the filament F, so that when they reach the grid G, once again they acquire enough energy to reach the plate P. The current thus begins to rise again. When V becomes equal to $2V_c$, a second inelastic collision occurs near the grid and the current drops again. Thus the second peak gives 2 V_c, . This process repeats as V is further increased.

The first peak occurs at a potential slightly less than V_c. This is because electrons are emitted from the filament with a finite velocity and hence with some energy. The true excitation potential V_e, is obtained by measuring the difference between two successive peaks.

Limitations : Atoms have more than one excitation potentials and also an unionisation potential. Therefore in an actual experiment the curve obtained is quite complicated and we cannot distinguish between which are excitation and which are unionisation potentials

Demonstration of the Existence of Discrete Energy Levels: The experiment shows that electrons transfer energy to the atoms in discrete amounts, and that they cannot excite atoms if their energy is less than eVc. Franck and Hertz demonstrated it directly by observing the spectrum of the gas during electron collisions. They showed that a particular spectral line does not appear until the electron energy reaches a *threshold* value. For example in the case of mercury vapour, they found that a minimum electron energy of 4.9 eV was required to obtain the 2536 A line of mercury; and a photon of 2536Å light has an energy of just 4.9 eV. This shows that discrete energy levels *do* exist in atom and the electrons of the atom can exist only in these levels.

Bohr's Correspondence Principle : The quantum theory gives results which are altogether different from those given by the classical theory in the microscopic world. Yet the two theories approach each other as the quantum number in question increases. This fact was pointed out by Bohr in 1923 who enunciated the following principle :

"The predictions of the quantum theory for the behaviour of any physical system must coincide with the predictions of the classical theory in the limit in which the quantum numbers specifying the state of the system become very large." This is known as 'Bohr's correspondence principle'. We establish it by means of an example.

According to classical theory, an electron moving in a circular orbit must emit radiation only of a particular frequency, namely, the frequency of revolution of the electron itself (or its integral multiples). According to quantum theory, on the other hand, radiation is emitted when the electron jumps from one orbit to the other and the frequency of radiation is determined by the difference in energy between the two orbits. Let us try to find a relation between the two frequencies.

The basic equations in the Bohr's theory of the hydrogen atom are the following :

$$\frac{mv^2}{r} = \frac{1}{4\pi \epsilon_0} \frac{e^2}{r^2}$$

$$\text{and} \qquad mvr = n\frac{h}{2\pi} \qquad n = 1, 2, 3, \ldots$$

where v is the velocity of the electron of mass m in the orbit of radius r. These equations give

$$v = \frac{nh}{2\pi mr}$$

$$\text{and} \qquad r = \frac{n^2h^2\epsilon_0}{\pi me^2}.$$

Hence the classical frequency of revolution of the electron is

$$f = \frac{\text{electron speed}}{\text{orbit circumference}} = \frac{v}{2\pi r}$$

$$= \frac{nh}{4\pi^2 mr^2} = \frac{nh}{4\pi^2 m\left(\frac{n^2h^2\,\epsilon_0}{\pi me^e}\right)^2}$$

$$= \frac{me^4}{4\,\epsilon_0^2\, n^3h^3} = \frac{me^4}{8\,\epsilon_0^2\, ch^3}\frac{2c}{n^3}$$

$$= \frac{2Rc}{n^3}, \qquad \ldots (i)$$

where $R\left(=\frac{me^4}{8\,\epsilon_0^2\, ch^3}\right)$ is the Rydberg constant. This is the frequency which must be radiated by the moving electron classically.

Now, the frequency radiated on the basis of quantum theory is given by

$$v = \frac{E_i - E_f}{h}$$

$$= \frac{me^4}{8\,\epsilon_0^2\, h^3}\left(\frac{1}{n_f^2} - \frac{1}{n_i^2}\right),$$

when the electron drops from the n_ith orbit to the n_fth orbit. Again, introducing R, we have

$$v = Rc\left(\frac{1}{n_f^2} - \frac{1}{n_i^2}\right).$$

This may be written in the form

$$v = Rc\frac{(n_i - n_f)(n_i + n_f)}{n_f^2\, n_i^2} \qquad \text{... } (ii)$$

Let us write $n_f = n$ and $n_i = n + 1$. Then the frequency of the emitted radiation for the transition $n + 1 \rightarrow n$ (so that $\Delta n = 1$) is given by

$$v = Rc\,\frac{2n+1}{n^2(n+1)^2}.$$

When n is very large, we can write $\dfrac{2n+1}{n^2(n+1)^2} \simeq \dfrac{2}{n^3}$.

Under this condition, the emitted frequency is

$$v = \frac{2Rc}{n^3} \qquad \text{...}(iii)$$

Equations (*i*) and (*iii*) yield

$$v = f$$

If we consider transitions $\Delta n = 2, 3, 4,$......we shall have

$$v = 2f, 3f, 4f, \ldots\ldots$$

Thus for very large quantum numbers n, the quantum-theory frequency v of the radiation is identical with the classical frequency of the revolution (or its harmonics) of the electron in the orbits. This is in accordance with Bohr's correspondence principle.

PROBLEMS

1. ***The speed of electron in the nth orbit of the hydrogen atom is calculated. If relativist correction is important for $\frac{v}{c} > 0.005$, find for which orbit this correction is necessary, ($e = 16 \times 10^{19}$ coul, $h = 6.63 \times 10^{-34}$ joule-see, $c = 3.0 \times 10^8$ m/sec and $\frac{1}{4\pi\epsilon_0} = 9 \times 10^9$ nt-m²/coul²).***

Solution: Let v be the speed of the electron (mass m, charge e) in the nth orbit whose radius is r, all in M.K.S. units. The charge on hydrogen nucleus is $+e$. As the centripetal force is provided by the electrostatic attraction, we have

$$\frac{mv^2}{r} = \frac{1}{4\pi\epsilon_0}\frac{e^2}{r^2}. \qquad \ldots (i)$$

Further, from Bohr's postulate, the angular momentum of the electron is an integral multiple of $h/2\pi$, that is,

$$mvr = \frac{nh}{2\pi}, \qquad n = 1, 2, 3, \ldots\ \ldots\ldots (ii)$$

Dividing eq. (*i*) by eq. (*ii*), we get

$$v = \frac{1}{4\pi\epsilon_0}\frac{2\pi e^2}{nh}.$$

Substituting the given values, we get

$$v = (9 \times 10^9 \text{ nt-m}^2/\text{coul}^2) \times \frac{2 \times 3.14 \times (1.6 \times 10^{-19} \text{ coul})^2}{n(6.63 \times 10^{-34} \text{ joule-sec})}$$

$$= \frac{1}{n}(2.18 \times 10^6) \text{ m/sec}.$$

This is the required value. Now

$$\frac{v}{c} = \frac{1}{n}\frac{2.18 \times 10^6}{3.0 \times 10^8} = \frac{0.0073}{n}.$$

Thus, for $n = 1, 2, 3, \ldots \ldots$ we have

$$\frac{v}{c} = 0.0073, 0.0036, 0.0024, \ldots \ldots$$

Hence the relativist correction is necessary for $n = 1$ orbit only.

2. *An expression for the radius of the electron orbit in the hydrogen atom in its nth state is found in the below-given way. What will be the approximate quantum number n for an electron in an orbit of radius 1 mm is also given below (Take required values from problem 1).*

Solution: The basic equations in the Bohr's theory of hydrogen atom are :

$$\frac{mv^2}{r} = \frac{1}{4\pi\epsilon_0}\frac{e^2}{r^2} \qquad \ldots (i)$$

and
$$mvr = \frac{nh}{2\pi}. \qquad \ldots (ii)$$

Squaring (ii) and dividing by (i). we get

$$r = 4\pi\epsilon_0 \frac{n^2h^2}{4\pi^2me^2}.$$

Substituting the given values, we get

$$r = \frac{1}{9 \times 10^9 \text{ nt-m}^2/\text{coul}^2}$$

$$\times \frac{n^2\left(6.63 \times 10^{-34} \text{ joule-sec}\right)^2}{4 \times (3.14)^2 \times \left(9.1 \times 10^{-31} \text{ kg}\right) \times \left(1.6 \times 10^{-19} \text{ coul}\right)^2}$$

$$= 0.53 \times 10^{-10}\, n^2 \text{ meter}$$

$$= 0.53\, n^2 \text{ Å}.$$

For $r = 1 \text{ mm} = 10^7$ Å, we have

$$10^7 = 0.53\, n^2$$

$$\therefore \quad n^2 = \frac{10^7}{0.53} = 18.87 \times 10^6$$

or $$n \simeq 4350.$$

3. *The time taken by the electron to traverse the first orbit in the hydrogen atom is calculated as follows. Electron mass and charge are 91 x 10–*^{31} *kg and 1.6 ×10–*^{19} *coul and h = 6.63 × 10*^{34} *J-s.*

Solution: The radius of the nth Bohr orbit is

$$r_n = 4\pi\epsilon_0 \frac{n^2h^2}{4\pi^2 me^2}$$

and the velocity of electron in the nth orbit is

$$v_n = \frac{1}{4\pi\epsilon_0} \frac{2\pi e^2}{nh}.$$

The time by the electron to traverse the nth orbit is therefore

$$T_n = \frac{2\pi\, e\, r_n}{v_n}$$

$$= 2\pi \left(4\pi\epsilon_0\right)^2 \frac{n^2h^2}{4\pi^2 me^2} \frac{nh}{2\pi e^2}$$

$$= \frac{4\epsilon_0^2 h^3 n^3}{me^4}.$$

For the first orbit, $n = 1$. Substituting the given values of h, m, e and ϵ_0=8.85 × 10^{-12} C^2/ N-m^2, we get

$$T_1 = \frac{4\left(8.85 \times 10^{-12}\right)^2 \left(6.63\times 10^{-34}\right)^3}{\left(9.1 \times 10^{-31}\right)\left(1.6 \times 10^{-19}\right)^4}$$

$$= 1.5 \times 10^{-16} \text{ s}.$$

4. *The average life-time of an electron in an excited state of hydrogen atom is about 10*$^{-8}$ *sec. Let us find as to how many*

revolutions does an electron in the n = 2 state make before dropping to the n = l state (R = 1.097 × 10⁷ meter⁻¹).

Solution: Let v be the velocity of electron (mass m, charge e) in an orbit of radius r. The basic equations are

$$\frac{mv^2}{r} = \frac{1}{4\pi \epsilon_0} \frac{e^2}{r^2}$$

and $$= mvr = n \frac{h}{2\pi}.$$

These equations give

$$v = \frac{nh}{2\pi mr}, \; r = \frac{n^2 h^2 \epsilon_o}{\pi me^2}.$$

The number of revolutions of the electron in the orbit per second is

$$f = \frac{v}{2\pi r} = \frac{2Rc}{n^3} \qquad \text{(as proved in Q, 8)}$$

For $n = 2$ state, we have

$$f = \frac{2 \times \left(1.097 \times 10^7 \text{ meter}^{-1}\right) \times \left(3 \times 10^8 \text{ meter/sec}\right)}{8}$$

$$= 8.2 \times 10^{14} \text{ per sec.}$$

Hence the number of revolutions of the electron in its life-time of 10^{-8} second.

$$= (8.2 \times 10^{14}) \times 10^{-8} = \mathbf{8.2 \times 10^6}.$$

5. ***The wavelength of the photon emitted when the hydrogen atom goes from n = 10 state to the ground state is found as follows.*** (R_H = **1.097 × 10⁻³ A⁻¹**).

Solution: Since the atom drops to the ground state (n = l), the photon emitted belongs to the Lyman series. The wavelengths of the spectral lines in this series are given by

$$\frac{1}{\lambda} = R_H\left(\frac{1}{1^2} - \frac{1}{n^2}\right); \qquad n = 2, 3, 4, \ldots\ldots$$

For $n = 10$, we have

$$\frac{1}{\lambda} = R_H\left(1 - \frac{1}{100}\right) = \frac{99}{100} R_H$$

or

$$\lambda = \frac{100}{99 R_H}$$

$$= \frac{100}{99 \times \left(1.097 \times 10^{-3}\ \text{Å}^{-1}\right)} = 921\ \text{Å}$$

6. *The wavelength of the first line of Balmer series is 6563 A. Rydberg constant is calculated follows.*

Solution: The wavelengths of the spectral lines of Balmer series are given by

$$\frac{1}{\lambda} = R_H\left(\frac{1}{2^2} - \frac{1}{n^2}\right); \quad n = 3, 4, 5, \ldots\ \ldots$$

For the first line ($\lambda = 6563 \times 10^{-10}$ m), $n = 3$.

$$\therefore \quad \frac{1}{6563 \times 10^{-10}\ \text{m}} = R_H\left(\frac{1}{2^2} - \frac{1}{3^2}\right) = \frac{5}{36} R_H$$

or

$$R_H = \frac{36}{5 \times \left(65 \times 10^{-10}\ \text{m}\right)} = 1.097 \times 10^7\ \text{m}^{-1}.$$

7. *The wavelengths of the first two lines of the Balmer series and the series limit is calculated as follows. ($h = 6.63 \times 10^{-34}$ joule-sec, $c = 3.0 \times 10^8$ meter/sec, $e = 1.6 \times 10^{-19}$ coul, m — 9.1×10^{-31} kg, $\epsilon_0 = 8.85 \times 10^{-12}$ coul²/nt-m²).*

Solution: Let us first calculate the Rydberg constant from the given data. We have

$$R = \frac{me^4}{8 \epsilon_0^2 ch^3}$$

$$= \frac{(9.1 \times 10^{31} \text{ kg}) \times (1.6 \times 10^{-19} \text{ coul})^4}{8 \times (8.85 \times 10^{12} \text{ coul}^2/\text{nt} - \text{m}^2)^2 \times (3.0 \times 10^8 \text{ m/sec}) \times (6.63 \times 10^{-34} \text{ joule-sec})^2}$$

$= 1.097 \times 10^7$ meter^{-1}.

Now, the wavelengths of the spectral lines of Balmer series are given by

$$\frac{1}{\lambda} = R\left(\frac{1}{2^2} - \frac{1}{n^2}\right); \qquad n = 3, 4, 5, \ldots \ldots$$

For the first line, $n = 3$.

$$\therefore \quad \frac{1}{\lambda} = R\left(\frac{1}{2^2} - \frac{1}{3^2}\right) = \frac{5R}{36}$$

$$\text{or } \lambda_1 = \frac{36}{5R} = \frac{36}{5 \times (1.097 \times 10^7 \text{ meter}^{-1})} = 6.563 \times 10^{-7} \text{ m}$$

$= 6563 \times 10^{-10}$ meter $= 6563$ Å.

For the second line, $n = 4$. We can see that $\lambda_2 = 4861$ Å.

For the series limit, n — ∞.

$$\therefore \quad \frac{1}{\lambda\infty} = \frac{R}{4}$$

$$\text{or} \quad \lambda\infty = \frac{4}{R} = \frac{4}{1.097 \times 10^7}$$

$= 3.646 \times 10^{-7}$ meter $= 3646$ Å.

8. *The first member of Balmer series of hydrogen has a wavelength of 6563 Å. Calculate the wavelength of the second member is calculated as follows.*

Solution: The wavelengths of the spectral lines of Balmer series are given by

$$\frac{1}{\lambda} = R\left(\frac{1}{2^2} - \frac{1}{n^2}\right); \qquad n = 3, 4, 5 \ldots \ldots$$

For the first member, $n = 3$.

$$\therefore \qquad \frac{1}{\lambda_1} = R\left(\frac{1}{2^2} - \frac{1}{3^2}\right) = \frac{5R}{36} \qquad \ldots (i)$$

For the second member, $n = 4$.

$$\therefore \qquad \frac{1}{\lambda_2} = R\left(\frac{1}{2^2} - \frac{1}{4^2}\right) = \frac{3R}{16} \qquad \ldots (ii)$$

Dividing eq. (*ii*) by (*i*), we get

$$\frac{\lambda_1}{\lambda_2} = \frac{3R}{16} \times \frac{36}{5R} = \frac{27}{20}.$$

$$\therefore \qquad \lambda_2 = \lambda_1 \times \frac{20}{27}$$

But $\lambda_1 = 6563$ Å (given).

$$\therefore \qquad \lambda_2 = (6563 \text{ Å}) \times \frac{20}{27} = 4861 \text{ Å}.$$

9. ***The first line of the Balmer series in the spectrum of hydrogen has a wavelength of 6563 Å. The wavelength of the first line of Lyman series in the same spectrum is calculated as follows.***

Solution: The wavelength of the spectral lines of hydrogen spectrum are given by

$$\frac{1}{\lambda} = R\left(\frac{1}{n_f^2} - \frac{1}{n_i^2}\right),$$

where R is Rydberg constant. For the first member of Balmer series, $n_i = 2$ and $nt = 3$.

$$\therefore \quad \frac{1}{\lambda_1} = R\left(\frac{1}{2^2} - \frac{1}{3^2}\right) = \frac{5R}{36}. \quad \ldots (i)$$

For the first member of the Lyman series, $nf = 1$ and $n_i = 2$.

$$\therefore \quad \frac{1}{\lambda_1'} = R\left(\frac{1}{1^2} - \frac{1}{2^2}\right) = \frac{3R}{4}. \quad \ldots (ii)$$

From (*i*) and (*ii*), we have

$$\frac{\lambda_1'}{\lambda_1} = \frac{5R}{36} \times \frac{4}{3R} = \frac{5}{27}.$$

$$\therefore \quad \lambda_1' = \frac{5}{27}\lambda_1 = \frac{5}{27} \times 6563 = 1215 \text{ Å}.$$

10. ***The series limit of Balmer series is at 3646 Å. The wavelength of the first member of this series is calculated as follows.***

Solution: The formula for Balmer series is

$$\frac{1}{\lambda} = R\left(\frac{1}{2^2} - \frac{1}{n^2}\right), \qquad n = 3, 4, 5, \ldots$$

For series limit $n = \infty$ and $\lambda = \lambda_{\infty}, = 3646$ Å.

$$\therefore \quad \frac{1}{3646 \text{ Å}} = \frac{R}{4}$$

$$\text{or} \quad R = \frac{4}{3646 \text{ Å}}.$$

Again, for the first member $n = 3$.

$$\therefore \quad \frac{1}{\lambda_1} = R\left(\frac{1}{2^2} - \frac{1}{3^2}\right) = \frac{5R}{36}$$

$$\text{or} \quad \lambda_1 = \frac{36}{5R} = \frac{36}{5 \times \dfrac{4}{3646 \text{ Å}}} = 6563 \text{ Å}.$$

11. ***An orange photon of wavelength 600 nm is emitted from an atom. The difference in energy in the two atomic states involved and the same result for the red 6563 Å line of hydrogen are found out as follows. (h = 6.63 × 10^{-34} joule-sec c = 3.0 × 10^8 meter/sec, 1 eV = l.6 × 10^{-19} joule).***

Solution: By Bohr's postulate, the emitted frequency is given by

$$v = \frac{\Delta E}{h},$$

where ΔE is the difference in energy. But $v = \frac{c}{\lambda}$ Therefore

$$\frac{c}{\lambda} = \frac{\Delta E}{h}$$

or $$\Delta E = \frac{hc}{\lambda}.$$

Here $$\lambda = 600 \text{ nm} = 600 \times 10^{-9} \text{ meter.}$$

$$\therefore \ \Delta E = \frac{\left(6.63 \times 10^{-34} \text{ joule-sec}\right) \times \left(3.0 \times 10^{2} \text{ meter / sec}\right)}{600 \times 10^{-9} \text{ meter}}$$

$$= 3.31 \times 10^{-19} \text{ joule}$$

$$= \frac{3.31 \times 10^{-19}}{1.6 \times 10^{-19}} = 2.07 \text{ eV.}$$

For λ = 6563 Å = 6563 × 10^{-10} meter, we can show that

$$\Delta E = 3.03 \times 10–^{19} \text{ joule} = 1.9 \text{ eV.}$$

12. ***The volume of energy required to remove an electron from a hydrogen atom in the ground state and also in a state with n = 8 is given as follows (ii) The corresponding energies of the singly-ionised helium atom is calculated. (m = 9.11 × 10^{-31} kg, e = 1.6 ×10^{-19} coul, h = 6.63 ×10^{-34} joule-sec, ϵ_0 = 8.85 × 10^{-12} coul^2/nt-m^2 and 1 eV = 1.6 × 10^{-19} joule).***

Solution: (*i*) The energy required to remove an electron from the hydrogen atom in the ground state (n = 1) to infinity (where the energy is zero) is numerically equal to the energy of the electron in the n = 1 orbit. Now, the energy of the electron in nth orbit of hydrogen (Z = 1) is given by

$$E_n = \frac{me^4}{8 \epsilon_0^2 h^2}\left(\frac{1}{n^2}\right)$$

For the ground orbit, n = 1.

$$\therefore \quad E_1 = -\frac{me^4}{8 \epsilon_0^2 h^2}$$

$$= -\frac{\left(9.11 \times 10^{31}\right) \times \left(1.6 \times 10^{-19}\right)^4}{8 \times \left(8.85 \times 10^{-12}\right)^2 \times \left(6.63 \times 10^{-34}\right)^2}$$

$$= -2.17 \times 10^{-18} \text{ joule}$$

$$= -\frac{2.17 \times 10^{-18}}{1.6 \times 10^{-19}} = -13.6 \text{ eV}.$$

Hence the energy required to remove the electron from n = 1 orbit to infinity is 13.6 eV.

This is known as the binding energy of the hydrogen atom.

Again, for n = 8 orbit,

$$E_8 = -\frac{me^4}{8 \epsilon_0^2 h^2}\left(\frac{1}{8^2}\right) = -\frac{E_1}{64}$$

$$= -\frac{13.6}{64} = -0.213 \text{ eV}.$$

Hence the energy required to remove an electron from n = 8 orbit to infinity is 0.213 eV.

(*ii*) The energy expression for He^+ (Z = 2) is

$$(E_n)_{He^+} = -\frac{mZ^2e^4}{8\epsilon_0^2 h^2}\left(\frac{1}{n^2}\right)$$

$$= -\frac{4me^4}{8\epsilon_0^2 h^2}\left(\frac{1}{n^2}\right) = -4\,(E_n)_H.$$

That is, the energies for He^+ are 4 times the corresponding energies for *H*. Thus the results are

4 × 13.6 =54.4 eV

4 × 0.213 =0.85 eV.

13. *The ground state energy of an electron in the hydrogen atom is —13.6 eV. The energy of (i) n = 3 state in He⁺, (ii) n = 2 state in Li⁺⁺ is computed as follows.*

Solution: The energy of an electron in the nth state of a hydrogen-like atom is given by

$$E_n = -\frac{mZ^2e^4}{8E_0^2h^2}\left(\frac{1}{n^2}\right)$$

$$= -\frac{RZ^2hc}{n^2}. \qquad \left[\therefore R = \frac{me^4}{8\epsilon_0^2 ch^3}\right]$$

For *H*, Z = 1, so that

$$(E_n)\,H = -\frac{Rhc}{n^2}.$$

When $n = 1$ (ground state), $(E_1)_H = -\,13.6$ eV (given).

$$\therefore \qquad Rhc = 13.6 \text{ eV}.$$

$$\text{Hence } (E_2)_H = \frac{Rhc}{4} = -\frac{13.6}{4} = -\,3.4 \text{ eV}$$

$$\text{and } (E_3)_H = \frac{Rhc}{9} = -\frac{13.6}{9} = -\,1.5 \text{ eV}.$$

(*i*) For He^+, Z = 2, so that

$$(E_3)_{He} = -\frac{4Rch}{9} = 4\,(E_3)_H = -\,6.0 \text{ eV}.$$

(*ii*) For Li^{++}, Z = 3, so that

$$(E_2)_{Li} = -\frac{9Rch}{4} = -30.6 \text{ eV}.$$

14. *The energy required to ionise hydrogen atom in the ground state, limit of Balmer series are calculated as follows. (Rydberg constant is 1.097 × 10^7 m^-1, h = 6.63 × 10^-34 joule-sec, c = 3.0 × 10^8 m/sec, 1 eV = 1.6 × 10^-19 joule).*

Solution: The energy required to ionise *i.e.* to remove an electron from the hydrogen atom in the ground state (n = 1) to infinity (where the energy is zero) is numerically equal to the energy of the electron in the n = 1 orbit. Now, the energy of the electron in nth orbit of hydrogen is given by

$$E_n = -\frac{me^4}{8\epsilon_0^2 h^2}\left(\frac{1}{n^2}\right)$$

$$= -\frac{Rhc}{n^2}. \qquad \left[\therefore R = \frac{me^4}{8\epsilon_0^2 ch^2}\right]$$

For the the ground (first) orbit, n = 1.

$$\therefore \quad E_1 = -Rhc$$

$$= -(1.097 \times 10^7 \text{ m}^{-1})(6.63 \times 10^{-34} \text{ J-s})(3 \times 10^8 \text{ m-s}^{-1})$$

$$= -21.8 \times 10^{-19} \text{ joule}$$

$$= -\frac{21.8 \times 10^{-19}}{1.6 \times 10^{-19}} = -13.6 \text{ eV}.$$

Hence the energy required to remove the electron from n = 1 orbit to infinity is 13.6 eV.

(*ii*) The Balmer series is represented by

$$\frac{1}{\lambda} = R\left(\frac{1}{2^2} - \frac{1}{n^2}\right); \qquad n = 3, 4, 5, \ldots\ldots$$

For the series limit $n = \infty$ so that

$$\frac{1}{\lambda} = \frac{R}{4}.$$

$$\therefore \qquad \lambda = \frac{4}{R} = \frac{4}{1.097 \times 10^{7}\ \text{m}^{-1}}$$

$$= 3.646 \times 10^{-7}\ \text{m} = 3646\ \text{Å}.$$

15. ***The wavelength of the first line of Balmer series of hydrogen is 6562.8 Å The ionisation potential and the first excitation potential of the hydrogen atom are calculated as follows. (h = 6.63 × 10⁻³⁴ J-s, c = 3 × 10⁸ m/sec).***

Solution: The wavelengths of the Balmer lines are given by

$$\frac{1}{\lambda} = R\left(\frac{1}{2^2} - \frac{1}{n^2}\right); \qquad n = 3, 4, 5, \ldots$$

For the first line $n = 3$ and $\lambda = 6562.8$ Å $= 6562.8 \times 10^{-10}$ m.

$$\therefore \qquad \frac{1}{6562.8 \times 10^{-10}} = R\left(\frac{1}{2^2} - \frac{1}{3^2}\right) = \frac{5R}{36}$$

Or
$$R = \frac{36}{5 \times 6562.8 \times 10^{-10}\ \text{m}}$$

$$= 1.097 \times 10^{7}\ \text{m}^{-1}$$

The ionisation potential of an atom is numerically equal to the (ionisation) energy required to remove an electron completely from the atom in the normal state. In hydrogen atom; the electron stays in the first orbit ($n = 1$). Hence the energy required to remove this electron to infinity (where the energy is considered to be zero) is numerically equal to the energy of the electron in the first orbit. We know that electron energy in hydrogen atom is given by

$$E_n = -\frac{Rhc}{n^2}$$

In the ground state, $n = 1$.

$$\therefore \quad E_1 = -Rhc$$

$$= -13.6\,\text{eV. (as obtained in the last problem).}$$

Hence the ionisation potential of the hydrogen atom is 13.6 volts.

(*ii*) The first excitation potential of the atom is the energy required to shift the electron from $n = 1$ to Now,

$$E_1 = -13.6 \text{ eV}$$

$$\text{and} \quad E_2 = \frac{E_1}{4} = -3.4 \text{ eV.} \qquad \left[\therefore E_n \propto \frac{1}{n^2}\right]$$

$$\therefore \quad E_1 \sim E_2 = 13.6 - 3.4 = 10.2 \text{ eV.}$$

Hence the first excitation potential is 10.2 volts.

16. ***A beam of electrons bombards a sample of hydrogen. Let us find as to through what minimum potential difference must the electrons have been accelerated if the first line of the Balmar series is to be emitted. The ionisation potential of hydrogen atom is 13.6 volts. The numbers of possible spectral lines can be expected if the atom finally attains the normal state is explained below.***

Solution: Normally the hydrogen atoms are in the ground state ($n = 1$). The first Balmar line is emitted when the atom returns from $n = 3$ to $n = 2$ state. Hence in order to emit this line the atoms must be first raised to the $n = 3$ state by electron bombardment. Therefore the energy of the bombarding electrons must be equal to the difference of energy between $n - 1$ and $n = 3$ states ($E_1 \sim E_3$).

The energy of hydrogen atom in the *nth* state is

$$E_n = -\frac{Rhc}{n^2}.$$

If E_1 is the energy in the ground ($n = 1$) state, then $E_1 = -Rhc$.

Thus $$E_n = \frac{E_1}{n^2}.$$

For $n = 3$;we have

$$E_3 = \frac{E_1}{9}.$$

$$\therefore \quad E_1 - E_3 = E_1 - \frac{E_1}{9} = \frac{8}{9} E_1.$$

But $E_1 = -13.6$ eV

$$\therefore \quad E_1 - E_3 = \frac{8}{9} \times (13.6) = 12.1 \text{ eV}.$$

Hence the bombarding electrons must be accelerated by 12'1 volts.

If the atom finally attains the normal state, the possible number of spectral lines is 3 (3→2, 3→l, 2→l).

17. ***A photon of energy 12.1 eV is completely absorbed by a hydrogen atom, originally in its ground state, so that the atom is excited. What is the quantum number of this state ? The ground state (negative) energy of hydrogen atom is 13.6 eV.***

Solution : Let the required quantum number be n.

Then $$E_1 - E_x = -12.1 \text{ eV}$$

$$E_1 - \frac{E_1}{n^2} = -12.1 \text{ eV}. \qquad \left[\because E_n = \frac{E_1}{n^2}\right]$$

But $$E_1 = -13.6 \text{ eV}.$$

$$\therefore \quad 1 - \frac{1}{n^2} = \frac{-12.1}{-13.6} = 0.89$$

or $$\frac{1}{n^2} = 1 - 0.89 = 0.11$$

or $$n^2 = \frac{1}{0.11} = 9$$

$$\therefore \quad n = 3.$$

18. ***A beam of monochromatic photons of energy 9 eV is incident on hydrogen gas all of whose atoms are in the ground state. It is found that the beam is fully transmitted without absorption. The reason is given ad follows. The ground state energy of an electron in the hydrogen atom is $E_1 = -13.6$ eV.***

Solution: The minimum energy that can be absorbed by ground-state hydrogen atom is $E_1 \sim E_2$, which would excite it to the next state ($n = 2$). Now,

$$E_1 \sim E_2 = E_1 \sim \frac{E_1}{4} \qquad \left[\because E_n = \frac{E_1}{n^2}\right]$$

$$= \frac{3}{4} E_1 = \frac{3}{4} \times 13.6 = 10.2 \text{ eV}.$$

Hence photons of energy 9 eV cannot be absorbed by hydrogen atoms.

19. ***The first two excitation potentials of atomic hydrogen in Franck-Hertz experiment are 10.2 and 12.09 volts. An energy level diagram and all possible transitions for emission and absorption along their wavelengths are shown as follows. ($h = 6.63 \times 10^{-34}$ joule-sec, $c = 3.0 \times 10^8$. m/sec, 1 eV = 1.6 $\times 10^{-19}$ joule).***

Solution: The energy level diagram, and the *three* possible transitions (*a*), (*b*) and (*c*) for emission are shown in the figure.

The frequency of radiation resulting from the transition (a) is given by

$$v_\alpha = \frac{E_2 - E_1}{h}$$

and the corresponding wave length is

$$\lambda_\alpha = \frac{hc}{(E_2 - E_1)}. \qquad [\because c = v\lambda]$$

Now, from the Fig., $E_2 - E_1 = 10.2$ eV $= 10.2 \times (1.6 \times 10^{-19})$ joule. Therefore

$$\lambda_a = \frac{(6.63 \times 10^{-34} \text{ joule-sec}) \times (3.0 \times 10^8 \text{ m/sec})}{(10.2 \times 1.6 \times 10^{-19} \text{ joule})}$$

$$= 1.216 \times 10^{-7} \text{ m} = 1216 \times 10^{-10} \text{ m} = 1216 \text{ Å}.$$

Similarly, $$\lambda_b = \frac{hc}{(E_3 - E_1)}$$

$$= \frac{(6.63 \times 10^{-34}) \times (3.0 \times 10^8)}{12.09 \times 1.6 \times 10^{-19}}$$

$$= 1.026 \times 10^{-7} \text{ m} = 1026 \times 10^{-10} \text{ m} = 1026 \text{ Å}.$$

Also $$\lambda_c = \frac{hc}{(E_3 - E_2)} = \frac{hc}{(E_3 - E_1) - (E_2 - E_1)}$$

$$= \frac{hc}{(12.09 - 10.2) \text{ eV}} = \frac{hc}{1.89 \text{ eV}}$$

$$= \frac{(6.63 \times 10^{-34}) \times (3.0 \times 10^8)}{(1.89 \times 1.6 \times 10^{-19})}$$

$$= 6.567 \times 10^{-7} \text{ m} = 6567 \times 10^{-10} \text{ m} = 6567 \text{ Å}.$$

In absorption, only the transitions starting from $n = 1$ shall be observed which correspond to 1216 Å and 1026 Å.

20. ***With Franck-Hertz type of experiment on sodium, the first spectral line to appear is the D-line, $\lambda = 5.89 \times 10^{-7}$ m. The first excitation potential of sodium is given below. Given : h = 6.63×10^{-34} joule-sec, c = 3×10^8 m/sec and leV = 1.6×10^{-19} joule.***

Solution: Let V volt be the excitation potential. Then the (excitation) energy imparted to the electron will be eV joule, where e coulomb is the charge on the electron. This energy is re-emitted as photon (radiation) when the electron returns to the normal state. If v be the frequency of the emitted radiation, the photon energy will be $h\nu$, Thus

$$eV = h\nu$$

or $$V = \frac{hv}{e}$$

If λ be the wavelength of the emitted radiation, then $v = c/\lambda$.

$$\therefore \quad V = \frac{hv}{e\lambda}$$

$$= \frac{(6.63 \times 10^{-34} \text{ J-s})(3 \times 10^8 \text{ ms}^{-1})}{(1.6 \times 10^{-19} \text{ C})(5.89 \times 10^{-7} \text{ m})}$$

$$= 2.1 \frac{\text{joule}}{\text{coulomb}} = 2.1 \text{ volt.}$$

21. ***The ionisation potential of an atom is 14.2 eV. The series limit in its absorption spectrum is calculated. Datas as in last problem.***

Solution : The series limit in the absorption spectrum of an atom corresponds to the energy which when absorbed by a ground state atom, ionises it. Thus, if V be the ionisation potential of an atom, the wavelength at the series limit is given by

$$\lambda = \frac{hc}{eV}$$

$$= \frac{(6.62 \times 10^{-34} \text{ joule-sec}) \times (3 \times 10^8 \text{ m-sec}^{-1})}{(14.2 \times 1.6 \times 10^{-19} \text{ joule})}$$

$$= 0.874 \times 10^{-7} \text{ m}$$

$$= 874 \text{ Å.}$$

22. ***The recoil speed of hydrogen atom after it emits a photon in going from n = 3 to n = 1 state is found out in following way. Electron mass is 9.11 × 10^{-31} kg and h — 6.626 × 10^{-34} J-s.***

Solution: The energy of the hydrogen atom in the *nth* state is given by

$$E_n = -\frac{Rhc}{n^2}$$

where R is Rydberg constant. From this, we get

$$E_1 - E_2 = -\frac{Rhc}{1^2} + \frac{Rhc}{3^2} = -\frac{8}{9}\ Rhc.$$

The energy of the emitted photon is

$$\Delta E = E_1 \sim E_3 = \frac{8}{9}\ Rhc.$$

The momentum of the photon is

$$p = \frac{\Delta E}{c} = \frac{8}{9}\ Rh.$$

By conservation of momentum, the recoil momentum of the hydrogen atom will be equal (and opposite) to the momentum of the emitted photon. The recoil speed of the atom is

$$v = \frac{\text{momentum}}{\text{mass}} = \frac{8}{9}\frac{Rh}{m_H}.$$

But $m_H = 1836\ m$, where m is electron mass.

$$\therefore \qquad v = \frac{8}{9}\frac{Rh}{(1836\ m)}.$$

Putting the given values of h, m and using $R = 1.097 \times 10^{-7}$ meter″^{1}, we get

$$v = \frac{8}{9}\frac{\left(1.097 \times 10^7\ \text{m}^{-1}\right)\left(6.626 \times 10^{-34}\ \text{J-s}\right)}{1836\left(9.11 \times 10^{-31}\ \text{kg}\right)}$$

$$= 3.86\ \text{m/s}.$$

23. ***A positronium atom is a system consisting of a positron and an electron. The reduced mass, the Rydberg constant and the wavelength of the first Balmer line for positronium are calculated as follows. (Given m = 9.1 × 10^{-3l} kg, R_H= l.09737 × 10^{-3} Å ~ ^{1} and H_α = 1.6563 Å).***

Solution: The positron has the same mass *m* as the electron and has equal but positive charge. The reduced mass of the electron-positron atom is therefore

$$\mu = \frac{(m)(m)}{m+m} = \frac{1}{2}m = 4.55 \times 10^{-31} \text{ kg},$$

while the reduced mass of electron in hydrogen is very nearly *m*.

The Rydberg constant $\left(\frac{\mu e^4}{8 \epsilon_0^2 ch^3}\right)$ for positronium is therefore half than that for hydrogen (with infinitely heavy nucleus). Thus

$$R_p = \frac{1}{2}R_H = 0.54868 \times 10^{-3} \text{ Å}^{-1}.$$

The wavelength of first Balmer line (H_α) for hydrogen atom is given by

$$\frac{1}{\lambda_H} = R_H\left(\frac{1}{2^2} - \frac{1}{3^2}\right),$$

while that for positronium atom is

$$\frac{1}{\lambda_p} = R_p\left(\frac{1}{2^2} - \frac{1}{3^2}\right),$$

Thus $$\frac{\lambda_p}{\lambda_H} = \frac{R_H}{R_p} = 2.$$

$$\therefore \qquad \lambda_p = 2\lambda_H = 2 \times 6563 = 13126 \text{ Å}$$

24. ***A μ meson (charge — e, mass = 207 m, where m is mass of electron) can be captured by a proton to form a hydrogen-like "mesic" atom. The radius of the first Bohr orbit, the binding energy, and the wavelength of the first time in the Lyman series for such an atom are calculated here. The mass of the proton is 1836 times the mass of electron. The radius***

of first Bohr orbit and the binding energy of hydrogen are 0.53 Å and 13.6 eV respectively. R_∞ = 109737 cm^{-1}.

Solution: The reduced mass of the system is

$$\mu = \frac{(207\,m)(1836\,m)}{207\,m + 1836\,m} = 186\ m.$$

From Bohr theory, the radius of the first orbit (n = 1) of a hydrogen-like atom for Z–1, is given by (taking finite mass of nucleus in consideration)

$$r_1 = 4\pi\epsilon_0 \frac{h^2}{4_n^2 \mu e^2} = 4\pi\epsilon_0 \frac{h^2}{4\pi^2 (186\,m)e^2}$$

$$= \frac{1}{186}\left(4\pi\epsilon_0 \frac{h^2}{4\pi^2 me^2}\right)$$

The quantity in the bracket is the first Bohr orbit of hydrogen atom which is 0.53 Å. Therefore

$$r_1 = \frac{1}{186} \times 0.53 \text{ Å} = 2.85 \times 10^{-3} \text{ Å}.$$

Again, from Bohr's theory, the ground-state energy for a hydrogen-like atom with Z = 1 is given by

$$E_1 = -\frac{\mu e^4}{8\epsilon_0^2 h^2} = -186 \frac{me^4}{8\epsilon_0^2 h^2}$$

$$= -186 \times 13.6 = -2530 \text{ eV}.$$

Hence the binding energy is 2530 eV.

The wavelengths of the Lyman lines are given by

$$\frac{1}{\lambda} = R_M \left(\frac{1}{1^2} - \frac{1}{n^2}\right), \qquad n = 2, 3, 4, \ldots$$

where R_m is she Rydberg constant for the mesic atom. For the first line, n = 2 so that

$$\lambda = \frac{4}{3R_M}.$$

Now, $$R_M = \frac{\mu e^4}{8 \epsilon_0^2 ch^3} \text{ and } R_\infty = \frac{me^4}{8 \epsilon_0^2 ch^3}, \text{ so that}$$

$$R_M = \frac{\mu}{m} R_\infty = 186\ R_\infty = 186 \times 109737 \text{ cm}^{-1}.$$

Hence $$\lambda = \frac{4}{3R_M} = \frac{4}{3 \times 186 \times 109737 \text{ cm}^{-1}}$$

$$= 6.52 \times 10^{-3} \text{ cm} = 6.52 \text{ Å}.$$

25. ***The wavelength of the first Balmer line (H_α) of hydrogen is 6562.8 Å. The wavelength of the same line for tritium ($_1H^3$) is calculated as follows.***

Solution: The wavelengths of the H_a line ($n = 3 \rightarrow n = 2$) of hydrogen ($_1H^1$) and tritium ($_1H'^3$) are given by

$$\frac{1}{\lambda_H} = R_H\left(\frac{1}{2^2} - \frac{1}{3^2}\right) = \frac{5}{36} R_H$$

and $$\frac{1}{\lambda_T} = R_T\left(\frac{1}{2^2} - \frac{1}{3^2}\right) = \frac{5}{36} R_T,$$

where R_H and R_I are Rydberg constants for hydrogen and tritium respectively (Z = 1 for both). From this, we get

$$\frac{\lambda_T}{\lambda_H} = \frac{R_H}{R_T}. \quad \text{... } (i)$$

We know that the Rydberg constant for an atom of mass M is given by

$$R_M = \frac{R_\infty}{1 + \frac{m}{M}},$$

where R_∞ is the Rydberg constant for *any* atom of infinitely heavy nucleus. Thus, eq. (*i*) gives

$$\frac{\lambda_T}{\lambda_H} = \frac{1+\dfrac{m}{M_T}}{1+\dfrac{m}{M_H}},$$

Now $M_T = 3M_H$ and $M_H = 1836\ m$. Thus

$$\frac{\lambda_T}{\lambda_H} = \frac{1+\dfrac{1}{3\times 1836}}{1+\dfrac{1}{1836}} = \frac{5509}{5511}.$$

But $\lambda_H = 6562.8$ Å.

$$\therefore \qquad \lambda_T = \frac{5509}{5511} \times 6562.8 \text{ Å} = 6560.4 \text{ Å}.$$

PROBLEMS IN THE C.G.S. SYSTEM

Note : The formulae deduced in this chapter belong to M.K.S. System. The values of h, *e*, *m*, *c* and *R* used in the above problems are also in the M.K.S. system. The formulae in the C.G.S. system are as follows : (They can be deduced exactly in the same manner as above by writing the force of electrostatic attraction between the electron and the nucleus as $\frac{e^2}{r^2}$ instead of $\left(\frac{1}{4\pi\,\epsilon_0}\frac{e^2}{r^2}\right)$:

$$r = \frac{n^2h^2}{4\pi^2me^2}$$

$$E = -\frac{2\pi^2me^4}{h^2}\left(\frac{1}{n^2}\right)$$

$$\frac{1}{\lambda} = \frac{2\pi^2me^4}{ch^3}\left(\frac{1}{n_f^2}-\frac{1}{n_i^2}\right)$$

$$= R\left(\frac{1}{n_f^2}-\frac{1}{n_i^2}\right),$$

where $R = 2\pi^2 me^4/ch^3$ (Rydberg constant).

In these formulae we shall use the following values :

$h = 663 \times 10^{-27}$ erg-sec, $e = 48 \times 10^{-10}$ esu, $m = 9.1 \times 10^{-28}$ gm, $c = 3.0 \times 10^{10}$ cm/sec, $R = 1.097 \times 10^5$ cm^{-1}, 1 eV = 1.6×10^{-12} erg, 1 Å = 10^{-8} cm.

1. *The hydrogen atom (i) radius of the first orbit, (ii) velocity of the electron in the ground state, (iii) Rydberg constant are calculated. (h = 6.63 × 10^{-27} erg-sec, m = 9.1 × 10^{-28} gm, e = 4.8 × 10^{-10} esu, c — 3.0 × 10^{10} em/sec and proton mass M_p = 1.67 ×10^{-24} gm)*

Solution : Let v cm/sec be the velocity of electron (mass m gm, charge e esu) in an orbit of radius r cm. The positive charge on the nucleus is Ze, where Z is atomic number. In case of hydrogen atom, Z = 1, so that positive charge on the nucleus is e. As the centripetal force is provided by the electrostatic attraction, we have

$$\frac{nv^2}{r} = \frac{e^2}{r^2}$$

or
$$mv^2 = \frac{e^2}{r}. \qquad \text{... } (i)$$

From Bohr's postulate, the angular momentum of the electron is given by

$$mvr = \frac{nh}{2\pi}, \qquad \text{... } (ii)$$

where n is an integer. Squaring (*ii*) and dividing by (*i*), we get

$$r = \frac{n^2h^2}{4\pi^2 me^2}. \qquad \text{... } (iii)$$

For the first orbit, $n = 1$.

$$\therefore \qquad r = \frac{h^2}{4\pi^2 me^2}$$

$$= \frac{(6.63 \times 19^{-27})^2}{4 \times (3.14)^2 \times (9.1 \times 10^{-28}) \times (4.8 \times 10^{-10})^2}$$

$$= 0.53 \times 10^{-8} \text{ cm} = 0.53 \text{ Å}.$$

(*ii*) Dividing eq. (*i*) by (*ii*), we obtain

$$v = \frac{2\pi e^2}{nh}.$$

For the ground state, $n = 1$. Therefore

$$v = \frac{2\pi e^2}{h}$$

$$= \frac{2 \times 3.14 \times (4.8 \times 10^{-10})^2}{6.63 \times 10^{-27}}$$

$$= 2.2 \times 10^8 \text{ cm/sec.}$$

(*iii*) The Rydberg constant is $R = \dfrac{2\pi^2 me^4}{ch^3}$. If we take the nuclear motion into account, then the Rydberg constant strictly for hydrogen is given by

$$R = \frac{2\pi^2 me^4}{ch^3}\left(\frac{1}{1+(m/M_p)}\right)$$

where M_p is the mass of proton (*i.e.* the hydrogen nucleus). Substituting the values

$$R_H = \frac{2 \times (3.14)^2 \times (9.1 \times 10^{-28}) \times (4.8 \times 10^{-10})^4}{(3.0 \times 10^{10}) \times (6.63 \times 10^{-27})}\left(\frac{1}{1+\dfrac{9.1 \times 10^{-28}}{1.67 \times 10^{24}}}\right)$$

$$= 1.096 \times 10^5 \text{ cm}^{-1}$$

2. The wavelength of the first member and the series limit for the Lyman series of hydrogen is calculated below, (h = 6.62 $\times 10^{-27}$ erg-sec ; m = 9.1 $\times 10^{-28}$ gm, e = 4.8 $\times 10^{-10}$ esu ; c = 3.0 $\times 10^{10}$ cm/sec).

Solution : According to Bohr's theory of hydrogen atom, the wavelengths of the Lyman series are given by

$$\frac{1}{\lambda} = R\left(\frac{1}{1^2} - \frac{1}{n^2}\right),$$

where R is the Rydberg's constant for hydrogen and n = 2, 3, 4, 5..

$$\text{Now} \qquad R = \frac{2\pi^2 me^4}{ch^3}$$

$$= \frac{2 \times (3.14)^2 \times \left(9.1 \times 10^{-28}\right) \times \left(4.8 \times 10^{-10}\right)^4}{\left(3.0 \times 10^{10}\right) \times \left(6.62 \times 10^{-27}\right)^3}$$

$$= 1.097 \times 10^5 \text{ cm}^{-1}.$$

$$\text{Hence} \qquad \frac{1}{\lambda} = 1.097 \times 10^5 \left(\frac{1}{1^2} - \frac{1}{n^2}\right) \text{ cm}^{-1}.$$

For the first line,n = 2.

$$\therefore \qquad \frac{1}{\lambda} = 1.097 \times 10^5 \left(\frac{1}{1^2} - \frac{1}{2^2}\right)$$

$$= 1.097 \times 10^5 \times \frac{3}{4} \text{ cm}^{-1}$$

$$\text{or} \qquad \lambda = \frac{4}{1.097 \times 10^5 \times 3} = 1.215 \times 10^{-5} \text{ cm}$$

$$= 1215 \times 10^{-8} \text{ cm} = 1215 \text{ Å}.$$

For the series limit, $n = \infty$.

$$\therefore \qquad \frac{1}{\lambda_\infty} = 1.097 \times 10^5 \left(\frac{1}{1^2} - \frac{1}{\infty}\right) = 1.097 \times 10^5$$

or λ_∞. $= 0.912 \times 10^{-5}$ cm $= 912 \times 10^{-8}$ cm = 912 Å.

3. *The change in energy when an electron in the hydrogen atom jumps from the first orbit to second is calculated ($e = 4.8 \times 10^{-10}$ esu, $h = 6.62 \times 10^{-27}$ erg-sec, $m = 9.1 \times 10^{-28}$ gm, 1 eV $= 1.6 \times 10^{-12}$. erg).*

Solution : The energy of an electron in nth orbit of hydrogen atom is given by

$$E_n = -\frac{2\pi^2 me^2}{n^2h^2}$$

The change in energy when electron jumps from $n = 1$ to $n = 2$ orbit is

$$\Delta E = E_2 - E_1 = -\frac{2\pi^2 me^4}{4h^2} - \left(-\frac{2\pi^2 me^4}{h^2}\right)$$

$$= \frac{3\pi^2 me^4}{2h^2}$$

$$= \frac{3 \times (3.14)^2 \times \left(9.1 \times 10^{-28}\right) \times \left(4.8 \times 10^{-10}\right)^4}{2 \times \left(6.62 \times 10^{-27}\right)^2}$$

$$= 1.63 \times 10^{-11} \text{ erg}$$

$$= \frac{1.63 \times 10^{-11}}{1.6 \times 10^{-12}} = 10.2 \text{ eV.}$$

13

Summary

According to Proton-Neutron Theory of Nuclear Constitution

(a) The number of protons gives the atomic number Z.

(b) The number of protons and neutrons together gives atomic mass numbers A.

(c) The number of neutron is (A - Z).

(d) The neutron/proton ratio increases with increasing mass number of atoms in order to ensure nuclear stability.

- The nuclear radius R is given by the expression

$$R = R_0 A^{1/3}$$

where R_0 has an average value of 1.4×10^{-5} m for all nuclei.

- The nuclear spin of different nuclei is given by

$$p_1 = I.\frac{h}{2\pi}$$

where the spin quantum number I has half integral value for isotopes of odd mass numbers and whole number values for those having even mass numbers.

- The nuclear magnetic moment is

$$\mu_I = g.\frac{h}{2\pi}\frac{e}{2M} = g\mu_n$$

- The nuclear magneton is

$$\mu_n = \frac{h}{2\pi}\frac{e}{2M} = 5.05 \times 10^{-27} \text{ A-m}$$

- Packing fraction is given by

$$f = \frac{M-A}{A} = \frac{\Delta m}{A}$$

$$= \frac{Mass\ defect}{Mass\ number}$$

It is defined as the mass defect per nucleon in the nucleus.

- Mass defect (or excess) is given by expression

$$\Delta m = (M - A)$$

where M is the measured mass and A the mass number of an atom.

- Nuclear binding energy is given by

$$\Delta E = 931 \times \Delta m \text{ MeV}$$

- The binding energy per nucleon is

$$= \frac{\Delta E}{A} = 931 \times \frac{\Delta m}{A} MeV$$

Its average value is 8 MeV/nucleon.

- A more accurate expression for true mass defect is

$$\Delta m = Z.m_H + (A - Z)m_n - M$$

Where

m_H = Mass of one hydrogen atom

m_n = Mass of one neutron

$$\Delta E = 931\ [Z.m_H + (A - Z)m_n - M] \text{MeV}$$

Nuclear Force : A part from long range coulombic forces between nuclear protons, there exist short-range attractive nuclear forces which keep various nucleons bound together in a nucleus. These forces are considered to arise from a continuous and rapid jumping back and forth pions between the two nucleons.

Liquid Drop Model : According to liquid drop model, a nucleus resembles a liquid drop in many respects. The short range nuclear forces play the part of surface tension and keep the nucleus in a spherical shape. The electrostatic repulsive forces play the role of disruptive forces which, tend to push the nucleon apart. The nuclear stability is determined by the balance of these two forces.

Nuclear Shell Model : Nuclear shell model considers the various nucleons to lead independent existence and move about in quantum shells just as do the extra nuclear electrons. Nuclear stability is due to complete and filled nuclear shells.

Two Body Problem and Nuclear Forces

1. When the two nucleons are bound together of the three possible bound states of a two nucleon system, di-neutron (nn), di-proton (pp) and deuteron (np), nature has provided us with only the deuteron and the other two are unstable.
2. When the two nucleons are in free state and one is made to impinge on the other i.e., the scattering processes.

In particle, it is not possible to make a neutron-target and therefore scattering experiments are limited only to neutron-proton (np) scattering and proton-proton (pp) scattering.

Properties of Deuteron

(a) Charge : +e

(b) Mass : 2.014735 amu

$= 3.34245 \times 10^{-27}$ kg

(c) Spin : l(In units of $\hbar$)

(d) Statistics : Bose-Einstein

(e) Radius : The root-mean-square value of deuteron radius is 2.1 fermi.

(f) Binding energy : 2.225 ± 0.003 MeV

The nuclear force binding the deuteron cannot be electrical because the nuetron carries no charge.

- In the ground state of deuteron, the proton and neutron spins are parallel (↑↑) and the deuteron is in a triplet spin state ($3s_1$)

1. Neutron is a half spin particle.
2. In the ground state of deuteron, the orbital angular momentum quantum number is zero (l = 0) and spin quantum number s = 1.

- Deuteron possesses only ground state and no excited state exist for the bound n-p system. Neutron and proton can form a stable combination (deuteron) only in the triplet state i.e. when the neutron and proton spins are parallel. The singlet state, i.e. a state of anti-parallel neutron-proton spins being unbound. Therefore the information about the nuclear force from the study of deuteron is limited only to triplet state.
- The existence of non-zero magnetic moment and electric quadrupole moment for deuteron suggests that at least a part of the neutron proton force acting in deuteron is non central i.e., tensor force.

Low Energy Scattering Experiments

Conclusions : (1) The neutron-proton (np) and the proton-proton (p-p) forces are the same when one takes into account the coulomb forces in p-p experiments. This shows that the nuclear force in all probability is charge independent.

(2) For the deuteron the state with spins anti-parallel (singlet state) is not bound by a margin of only about 50 KeV i.e., it is unbound by about 50 KeV. This shows that the nuclear force depends on the spin orientation of the two nucleons.

Comparative Study of Shell Model and the Liquid Drop Model : In the shell model, it has been assumed that each nucleon moves in its orbit with the nucleus, independently of all other nucleons. The orbit is determined by a potential energy function V(r). Each nucleon is regarded as an independent particle and the interaction between nucleons is considered to be a small perturbation on their action between a nucleon and the potential field. Thus, the interaction between the orbital nucleon and the rest is very weak.

- In the liquid drop model on the other hand, the nucleons are considered to interact strongly with each other so that the collective motions are possible.

Collective Model : The shell model has been successful in explaining a number of nuclear properties but it fails to explain.

1. The large electric quadrupole moments and spherical shapes which many nuclei possess.
2. The ground states of the odd nuclei in the range $150 \leq A \leq 190$ and at $A \geq 220$.
3. The magnetic moments of some other nuclei where the deviations are not so marked.
4. The excited states of even-even nuclei, the probabilities of radiative transition and nuclear coulomb excitations. The existence of large electric quadrupole moments for certain nuclei is a clear indication that the nuclear surface is no longer spherical but deformed the simplest such deformed surface is spheroid.

The deformation of the nucleus is attributed to the polarizing action of one or more loosely bound nucleons on the remaining nucleus. The nucleons move in a potential which is not spherically symmetrical.

∴ We have two types of motions :

1. The motion of the entire nucleus with nucleons occupied in an ellipsoidal box that might rotate or deform itself by vibrations.
2. The motion of the nucleons inside the box. These two types of motions are coupled to each other. This forms the basis of collective models.

There are two different types of models.

In the first one, the nucleus is separated into a core and extra core nucleons. The core is treated macroscopically as deformable drop of nuclear 'liquid' in interaction with the few extra nucleons in an unfilled shell. This model is some times called the Collective Model.

In the second case, the shell model potential is assumed non-spherical. The energies of the single particles in the non-spherical potentials are calculated and the distortion which gives minimum energy is taken as the actual distortion. This model is known as unified model.

Optical Model : A modification of the potential-well model in which the potential is made complex to account both for elastic scattering and for nuclear reactions. The latter effectively remove nucleons or clusters from the beam of incident particles.

- For basic principles of particle detectors-(various types) (Experimental design).

Natural Radioactivity : The phenomenon of spontaneous emission of powerful radiations exhibited by heavy elements is called radioactivity.

- Radioactive radiations consists of α-rays, β-rays and γ-rays.
- The two laws of radioactive disintegration are—

1. Atoms of all radioactive elements undergo spontaneous disintegration to form fresh radioactive products with the emission of α-, β- and γ-rays.

2. The rate of radioactive disintegration is not affected by environmental factors but depends on the number of the atoms of the original kind present at any time. The decay is given by

$$N = N_0 e^{-\lambda t}$$

Also, $$M = M_0 e^{-\lambda t}$$

- The decay constant X may be defined as the reciprocal of the time during which the number of atoms of a radioactive substance falls to 37 per cent of its original value.
- Half life (T) of a radioactive element is equal to the time during which a given amount of that element is reduced by disintegration to half its initial amount

$$T = \frac{0.693}{\lambda}$$

- The average life expectancy of radioactive atoms is equal to the reciprocal of-the decay constant

$$l = \frac{1}{\lambda}$$

- The rate of disintegration or activity of a radioactive material is given by

$$-\frac{dN}{dt} = lt$$

- Curie (c) is the unit of activity i.e., rate of disintegration. One curie is that amount of radioactive material which gives 3.7×10^{10} disintegrations per second.
- Rutherford (rd) is also the unit of activity and is equal to that quantity of radioactive material which gives 10^6 disintegration per second.
- For radioactive equilibrium

$$\left(\frac{dN}{dt}\right)_{Parent} = \left(\frac{dN}{dt}\right)_{Daughter}$$

- Radioactive exposure is measured in terms of roentgen (R) and one roentgen is that quantity of X- or γ-rays which under standard pressure and temperature produces 2.08×10^{15} ion-pairs per m^3 of dry air.

Artificial Radioactivity

- Artificial radioactivity was discovered by Curie-Joliots when they bombarded aluminium with α-particles from polonium.
- Radiosodium was produced by bombarding Na with deuterons obtained from a cyclotron.
- Transuranium elements have been produced by bombarding uranium with fast deuterons and neutrons and accelerated α-particles.
- Radioisotopes have found the following uses :

1. As tracer elements
2. In diagnosis and understanding of many life processes and diseases which cannot be handled by conventional methods.
3. In research and industry.
4. For dating purposes in geology.

Nuclear Reactions

- The study of nuclear reactions involves the measurement of the following quantities :

1. Intensity, energy and identity of the incident beam of particles.
2. The number of particles which are emitted from the target per unit time.
3. Angular distribution of the reaction products.
4. The energy and identity of the emitted particles and of the residual nucleus.
5. Induced activity of the product nucleus if any.

- The overall energy liberated or absorbed in a nuclear disintegration is called nuclear reaction energy Q.
- Those nuclear reactions in which energy is liberated are called exothermic or exoergic. They have positive Q-value.
- Those reactions in which energy is absorbed are called endothermic or endoergic. They have a negative Q-value.

Considerations

1. When the energy of the incident particle lies below about 10 MeV, the nucleus as a whole acts as one entity.
2. When the energy of the incident particle is of the order of few hundreds of MeV, interactions with local clusters of nucleons predominate.
3. At bombarding energy of several hundred MeV, meson production predominates and nuclear structure becomes of little importance.

- In general, the Q-value of nuclear reaction is given by

$$Q = (M_0 + M_1) - (M_2 + M_3)$$
$$= (E_2 + E_3) - E_1$$

Classification of Nuclear Reaction

1. Low energy reactions ($0 < E^i < 100$ eV)
2. Moderate energy reactions (1keV $< E^i <$ 500 keV)
3. High energy reactions (0.5 MeV $< E^i <$ 10 MeV)
4. Very high energy reactions (10 MeV $< E^i <$ 50 MeV)
5. Ultra high energy reactions ($E^i >$ 50 MeV) Target nuclei with $A < 25$ = Light nuclei

   $25 < A < 80$ = Intermediate nuclei

   $80 < A < 250$ = Heavy nuclei

(a) *Proton Induced Nuclear Reaction* : Those reactions which include (p, α), (p, n), (p, γ) and (p, d) reactions.

(i) *(p, α) Reaction* : In general

$${}_{Z}X^{A} + {}_{1}H^{1} \rightarrow [{}_{Z+1}C^{A+1}] \rightarrow {}_{Z-1}Y^{A-3} + {}_{2}He^{4}$$

(ii) *(p, n) Reactions* : In general

$$_ZX^A + {}_1H^1 \rightarrow [_{Z+1}C^{A+1}] \rightarrow {}_{Z+1}Y^A + {}_0n^1$$

(iii) *(p, γ) Reactions* R The general scheme of such reaction is

$$_ZX^A + {}_1H^1 \rightarrow [_{Z+1}C^{A+1}] \rightarrow {}_{Z+1}Y^{A+1} + \gamma$$

(iv) *(p, d) Reaction* : The general scheme of these reaction is

$$_ZX^A + {}_1H^1 \rightarrow [_{Z+1}C^{A+1}] \rightarrow {}_{Z-1}Y^{A-1} + {}_1H^2$$

(b) *Deuteron induced Nuclear Reactions* : These reactions include (d, α), (d, p) and (d, n) types of reactions.

On general scheme, all these reactions may be written as

$$_ZX^A + {}_1H^2 \rightarrow [_{Z+1}C^{A+2}]$$

$$\rightarrow {}_{Z-A}Y^{A-2} + {}_2He^4 \rightarrow (d, \alpha) \text{ reaction}$$

$$\rightarrow {}_ZX^{A+2} + {}_1H^1 \rightarrow (d, p) \text{ reaction}$$

$$\rightarrow {}_{Z+1}Y^{A+1} + {}_0n^1 \rightarrow (d, n) \text{ reaction}$$

(c) *Neutron induced Reactions* : Neutron induced reactions include the following types—

(i) (n, α), (ii) (n, p), (iii) (n, γ) and (n, 2n)

(i) *(n, α) Reactions* : The general scheme of (n, α) reactions is

$$_ZX^A + {}_0n^1 \text{®} [_ZC^{A+1}] \text{®} {}_{A-2}Y^{A-3} + {}_2He^4$$

(ii) *(n, p) Reactions* : General scheme

$$_ZX^A + {}_0n^1 \rightarrow [_ZC^{A+1}] \rightarrow {}_{Z-1}Y^{A-1} + 2{}_0n^1$$

(iii) *(n, 2n) Reactions* : General scheme

$$_ZX^A + {}_0n^1 \rightarrow [_ZC^{A+1}] \rightarrow {}_ZX^{A-1} + 2{}_0n^1$$

- The Q-value of (n, 2n) reaction is always negative and therefore can be produced only by the fast neutrons.
- Another type of reaction induced by neutrons is the so called radioactive capture of neutrons, in which photons (γ-rays) are emitted out and the product nucleus is an

isotope of the target nucleus with a mass number one unit greater.

- General scheme of such reaction is

$$_{Z}X^{A} + {}_{0}n^{1} \rightarrow [_{Z}C^{A+1}] \rightarrow {}_{Z}X^{A+1} + \gamma$$

Photo-disintegration (Nuclear Reactions Induced by Photons or Y-rays) : Photo-disintegration is a process in which disintegration is produced by bombardment through high energy photons.

- Photons possess zero rest mass energy and therefore are capable of delivering only their kinetic energy and for the ejection of a particle from the target nucleus, the kinetic energy of the photon must exceed or be at least equal to the binding energy of the target nucleus.
- Photo-disintegration reactions are endoergic with threshold energies of the order of 10 MeV.
- Examples of photo-disintegration reactions

$$_{1}H^{2} + \gamma \rightarrow [_{1}H^{2}] \rightarrow {}_{1}H^{1} + {}_{0}n^{1}$$

$$_{4}Be^{9} + \gamma \rightarrow [_{4}Be^{9}] \rightarrow {}_{4}Be^{8} + {}_{0}n^{1}$$

$$_{15}P^{31} + \gamma \rightarrow [_{15}P^{31}] \rightarrow {}_{15}P^{30} + {}_{0}n^{1}$$

These are the examples of $[\gamma, n]$ reactions.

Nuclear Fission and Fusion

- The division of a nuclide into two approximately equal parts when hit by neutron is called nuclear fission.
- The fission fragments are initially radioactive and ultimately become stable after emitting a few β-particles,
- Fissile nuclides are those which undergo fission with neutrons of energy ranging from almost zero upwards. Examples are—U^{235}, U^{233} and Pu^{239}.

  Fissionable nuclides are those which have fission threshold at 1 MeV. Examples are— U^{238} and Th^{238}.

- Fission energy per U^{235} nuclide is nearly 200 MeV and per nucleon = $\frac{200}{235}$ = 0.85 MeV.
- Nuclear fission means fusing of two lighter nuclei into one stable and heavier nuclide. The energy released is called fusion energy and its value is nearly 6.7 MeV/nucleon.

Nuclear Energy Sources

- Energy released by complete fission of 1kg of U^{235} is

$$= 8.2 \times 10^{13} \text{ J/kg}$$
$$= 8.2 \times 10^{10} \text{ J/g}$$

- Rate of nuclear fission of U^{235} necessary to produce a power of one watt = 3.1×10^{10} fissions/second.
- A chain reaction is that fission process in which number of neutrons keeps on multiplying during fission till whole of the fissile material is disintegrated.
- The multiplying factor or reproduction factor of a chain reaction is

$$k = \frac{\text{No. of neutrons in any particular generation}}{\text{No. of neutrons in the preceding generation}}$$

  Chain reaction will continue if $k > 1$.

- Critical size of a fissile material is that for which production of neutrons by fission is equal to their loss by leakage and-non-fission capture. For this size, $k = 1$.
- For super critical size, $k > 1$ and for subcritical size, $k < 1$.
- A chain reaction in natural uranium is possible only if moderator is used whose purpose is to thermalise the fast fission neutrons.
- Power of a nuclear reactor is

$$P = 32 \times 10^{-11} \phi\sigma N V \text{watt}$$

Elementary Particles

Particle Interaction : To understand the behaviour of elementary particles, we must know the nature of the interactions between them, which are categorised into four classes as follows—

Gravitational Interaction : It can be expressed as

$$F = G\frac{m_1 m_2}{r}$$

F → force acting between the two particles of mass m_1 and m_2 at distance r apart and G is gravitational constant. Nature : Attractive and source : Inertia of the particle.

- The direction of action is along the line joining the two particles and it shows $\frac{1}{r^2}$ dependence on the distance between them.

Magnitude of interaction for nucleons at a separation distance of nuclear diameter (10^{-15} metre) is given as—

$$F_g = G\frac{m_1 m_2}{r^2}$$

$$= 6.7\times10^{-11}\frac{\left(1.7\times10^{-27}\right)^2}{\left(10^{-15}\right)^2}$$

$$= 2 \times 10^{-34} \text{ Newton}$$

The force does not depend upon the velocity and orientation of masses and other parameters.

The force is propagated through a particle known as graviton, not detected so far.

Electromagnetic Interaction : It depends on the nature and magnitude of charges of the particles. The interaction reduces to its simplest form when particles are at rest and is known as electrostatic force and expressed as—

$$F_e = k\frac{q_1 q_2}{r^2}$$

Fe → Force acting between the two particles having charges q_1 and q_2 and kept at a distance r apart.

Constant k, depends on space in which charges are placed.

- Direction of force is again along the joining line.
- Magnitude for two protons is

$$F = 9\times10^{9}\frac{\left(1.6\times10^{-19}\right)}{\left(10^{-15}\right)^{2}}\sim 30 \text{ Newton}$$

This is 10^{35} times as large as gravitational interaction.

The agent through which the field propagates is photon.

Strong Interaction : The forces which hold the nucleons together in the nuclei of the atoms are the examples of strong interaction.

- The forces are short range (1.3×10^{-15} metre) and charge independent.

The forces do not depend upon the relative orientations of nucleons. It is found that strong interaction is 1000 times stronger than electromagnetic interaction. The agent through which the interaction propagates is pion or kaon.

Weak Interaction : The interaction in β-decay process is the example of weak interaction.

- There are several such decay process in which weak interaction is involved. The weak interaction differs from others in the sense that it is never attractive.

The range of weak interaction is very short, even less than 10^{-15} m.

The weak interaction is about 10^{-13} times of strong interaction.

Classification of Elementary Particles : The elementary particles are classified into four groups namely, massless bosons, leptons, mesons and baryons.

The Parameters : The systematic study of elementary particles requires the knowledge of characteristic variables associated with the particles. The characteristic variables are known as parameters and they are :

Mass : The elementary particles differ from each other in respect of their masses. The mass of a particle is measured in terms of the energy it takes to produce it. This is generally expressed in eV.

Particle, Antiparticle

- Three fundamental particles viz. electron, proton and photon.
- The first particle is negatively charged, second is positively charged and third one is neutral.

  The development of Dirac theory led the negative energy states for the electrons which in turn gave birth to a new particle, known as positron.
- This particle has the same mass as electron but it has got positive charge. The positron is known antiparticle of electron. Similarly other particles are said to have their antiparticles.

  The antiparticles are denoted by the same symbol but with a bar over the symbol.
- e.g., proton is denoted by p and antiproton by $\overline{p}$
- Most of the particles and antiparticles have the same mass and spin but opposite charge and strangeness number. They also differ in respect of the sign of their magnetic moments and intrinsic parity (only fermions).

Quantum Numbers

(a) *The Nucleon Number* : The charge independence of nuclear interactions suggests that neutron and protons are the two states of single entity, the nucleon. It is also known that neutron converts into proton and vice versa and free proton is stable. For this reason nucleon number N is defined as

N = (Number of nucleons)—(Number of anti-nucleons)

N $\rightarrow$ Quantum number, and remains conserved in decay processes.

(b) *The Lepton Number* : The introduction of nucleon number makes one to think about a similar quantum number for electron which is also a fairly stable particle. Such type of quantum number is defined for a family to which the electron belong. The family is called lepton and includes muon, electron and neutrions.

The lepton number is written as :

L = (Number of leptons) - (Number of anti-leptons)

This is again a quantum number and remains constant. The conservation of lepton number makes the decay schemes of pions and muons to run as—

$\pi^+ \rightarrow \mu^+ + v_\mu$

$\pi^- \rightarrow \mu^- + \overline{v}_\mu$

$\mu^+ \rightarrow e^+ + \overline{v}_\mu + v_e$

$\mu^- \rightarrow e^- + v_\mu + v_e$

(c) (i) *The Baryon Numbers* : The behaviour of nucleons, leptons and hyperons is collectively expressed by a new quantum number known as baryon number B. Baryon number B assumes the values as follows—

B = + 1, for nucleons and hyperons (baryons)

B = –1, for anti-baryons

B = 0, for all other particles

The baryon number is conserved in all interactions.

(ii) *The Muon Number M* : The M = 1 is assigned to negative muon and to the μ-neutrino and M = –1 to their anti-particles; all other particles have M = 0. The significance of M is that in every process of whatever kind, the number M remains constant.

- The pion decay, muon decay and pair production illustrate the conservation of lepton number and muon number. The process are—

(i) $\pi^o \rightarrow \mu^- + \overline{\nu}_\mu$

$$L : 0 \rightarrow 1 + (-1) \qquad \delta L = 0$$

$$M : 0 \rightarrow +1 + (-1) \qquad \delta M = 0$$

(ii) $\mu^- \rightarrow e^- + \nu_\mu + \overline{\nu}_\mu$

$$L: +1 \rightarrow 1 + 1 - 1 \qquad \delta L = 0$$

$$M: -1 \rightarrow 0 + (-1) + 0 \qquad \delta M = 0$$

$$\therefore \quad \gamma = e^- + e^+$$

$$L : 0 \rightarrow 1 + (-1) \qquad \delta L = 0$$

$$M : 0 \rightarrow 0 + 0 \qquad \delta M = 0$$

Muon number can be included in lepton number but conservation laws applies separately for both. As strong interaction do not produce leptons, lepton conservation do not apply to strong interactions.

(d) *Spin Quantum Number and Statistics* : The intrinsic spin of two particle is expressed by the letter S and is known as spin quantum number.

- The spin quantum number assumes odd half integral values for fermions (particles obey F.D. statistics) and integral values for bosons (particles obey B.E. statistics).
- Fermions are subjected to the conservation principle i.e., whenever they are produced with some antiparticles.
- Alternately, we say that lepton number and baryon number remains constant.
- No such conservation principle is applicable for bosons i.e., bosons are produced without any antiparticle, provided charge, momentum etc. remain conserved.

(e) *Iso-spin Quantum Number* : We know, neutron and proton has get almost the same mass, same intrinsic spin; also that nuclear forces are charge independent which inferences that the two particles are different charge manifestation of the same entity the nucleon.

- Thus, the particles are grouped like isotopic grouping of the elements. The charge is treated as variable for different state of nucleon and this variable is known with the help of a new quantum number known as isotopic or iso-baric quantum numbers.
- But since neutron and protons are not isotopes, iso-spin quantum number is used.
- The new variable plays similar role as spin quantum numbers for a particle. The similarity between the two vectors suggested the scientists to assign a value I

$$= \frac{1}{2} \text{ for nucleon and } I = 1 \text{ for pion.}$$

So, Isospin vector (I_3)

$$\left.\begin{array}{l} I_3 = \frac{1}{2} \text{ for the proton state} \\ I_3 = -\frac{1}{2} \text{ for the neutron state} \end{array}\right\} \text{for nucleon}$$

$$\left.\begin{array}{l} I_3 = +1 \text{ for the } \pi^+ \text{state} \\ I_3 = 0 \text{ for the } \pi^0 \text{state} \\ I_3 = -1 \text{ for the } \pi^- \text{state} \end{array}\right\} \text{for pion}$$

(f) *Strangeness* : Introduced by Gellmann and Nishijima for observing the strange behaviour of k-mesons and hyperons, known as strange particles.

- The strangeness quantum number is assumed zero for the particles which are not strange and it has non zero value for strange particles. The strangeness quantum numbers is defined as

$$S = 2\Delta \overline{Q}$$

$\Delta \overline{Q} \rightarrow$ Shift in mean charge centre of strange particle from reference origin of nucleon multiplet.

$$S = 2\Delta \overline{Q}$$

- $S = -1$ [for Λ° and Σ hyperons]

$= -2$ [for Ξ doublet]

$= -3$ [for Ω - doublet]

- The charge Q of particle is

$$Q = I_3 + \frac{B+S}{2}$$

(g) *Hyper Charge* : It is expressed as sum of baryon and strangeness number

$$Y = B + S$$

- By this definition pion have hyper charge zero kaon and nucleons have hypercharge, Y =1.
- Also defined as the twice the average charge of the members of given multiplet.
- For the pion triplet (π^+, π^0, π^-) the average charge is zero and hence all the three have zero hyper charge.
- For kaons multiplets; (K^+, K^0) and ($\overline{K}^-$, $\overline{K}^0$), hypercharge is +1 and –1 respec-tively.
- Mathematically

$$Y = 2(Q - I_3)$$

- For particle and antiparticle hypercharge will be numerically equally and opposite in sign or zero.
- As baryon number is conserved, the hyper-charge is conserved in all strong and electromagnetic interactions.

Parity

- If we reflect the position co-ordinates of the system to obtain a mirror image, then there are two possibilities, namely; the function may change sign and may not. If the function changes sign it is said to possess odd parity and if it does not it is said to possess even parity.

  The reflection operation of position coordinates is known as parity operation.

  Parity is related to the symmetry properties of fundamental physical phenomena. The reflection symmetry principle of physical laws is equivalent to the following statement. "The parity of a closed system cannot change". Parity is found to conserve in strong and electromagnetic interactions but it is not conserve in weak interaction such as B-decay.

- Electron, proton and neutron are assigned even parity.
- The parity of Λ, Σ and Ξ hyperons is also even.
- The pions, K mesons and η° mesons have odd parity.

Charge Conjugation : The operation of changing every particle by its antiparticle is known as charge conjugation (C). It is a symmetry operation.

- Charge conjugation invariance means that nothing is changed by charge conjugation.
- The parity is not conserved in β decay but it appeared plausible that β interaction is invariant under the combined operations (CP = G) of charge conjugation and parity operations.
- CP symmetry is violated in weak interactions.

Time Reversal : A third symmetry operation which has been regarded as invariant in time reversal (ò).

- The time reversal invariance means that a positive sense of time cannot be distinguished from negative sense of time i.e., if we replace t by -t, all velocity and momentum vectors are reversed.
- The three symmetry operations constitute an important theorem in particle physics and is known as CPT theorem.
- The theorem states that all kinds of interactions (weak and strong) are invariant under the combined operations of C, P and T taken in any order.
- Melvin has alternately stated the theorem as—

"The processes seen directly in matter look exactly like those on film of processes in anti matter taken through a mirror and run off backward".

- The operation (CP = G) is conserved in beta decay. The CPT theorem requires that β

Tabulation of Some Elementary Particles

Charge	*Spin geness B + S*	*Stran- Y= I′ I_3*	*Hyper Charge ← →*	*Isospin I T_3*	*Particle (Symbol)*
Photon					
γ	0	1	—	—	—
Leptons					
Electron-neutrino ν_e	0	$\frac{1}{2}$	—	—	—
Muon-neutrino νμ	0	$\frac{1}{2}$	—	—	—

Charge		*Spin geness B + S*	*Stran- Y= I′ I_3*	*Hyper Charge ← →*	*Isospin I T_3*	*Particle (Symbol)*
Electron (e^-)		–1	$\frac{1}{2}$	—	—	—
Muon (μ^-)		–1	$\frac{1}{2}$	—	—	—
Hadrons					**Mesons**	
π^+	+1	0	0	0	1	1
π^-	–1	0	0	0	1	–1
π^0	0	0	0	0	1	0
K^+	+1	0	+1	+1	$\frac{1}{2}$	$+\frac{1}{2}$
K^-	0	0	+1	–1	$\frac{1}{2}$	$-\frac{1}{2}$
K^0	0	0	+1	+1	$\frac{1}{2}$	$+\frac{1}{2}$
$\overline{K}^0$	0	0	–1	–1	$\frac{1}{2}$	$-\frac{1}{2}$
η^0	0	0	0	0	0	0

Baryons						
Proton (P)	1	$\frac{1}{2}$	0	1	$\frac{1}{2}$	$\frac{1}{2}$
Neutron (n)	0	$\frac{1}{2}$	0	1	$\frac{1}{2}$	$-\frac{1}{2}$
Lambda (Λ^0)	0	$\frac{1}{2}$	–1	0	0	0
Sigma(Σ^+)	+1	$\frac{1}{2}$	–1	0	1	1
Sigma(Σ^0)	0	$\frac{1}{2}$	–1	0	1	0
Sigma(Σ^-)	–1	$\frac{1}{2}$	–1	0	1	–1
$\chi_i(\Xi^0)$	0	$\frac{1}{2}$	–2	–1	$\frac{1}{2}$	$\frac{1}{2}$
$\chi_i(\Xi^-)$	–1	$\frac{1}{2}$	–2	–1	$\frac{1}{2}$	$-\frac{1}{2}$
Ω^-	–1	$\frac{2}{3}$	–3	–2	0	0

decay is conserved under time reversal. If in same weak interaction CP interaction breaks down, then CPT invariance requires the break down of time reversal invariance.

Quark Flavours and Colours : According to this model each hadronic particle is made up of basic units called quarks. In this model we need only three types of quarks, namely u, d and s. The charge of these are $\frac{2}{3}e, -\frac{1}{3}e$ and $-\frac{1}{3}e$ respectively. For example, a proton is made up of three quarks namely uud, and a neutron is made of udd. Further, a π^* meson is a bound state of u and $\bar{d}$ (the antiquark of d). In this way we find all the baryons could be derived. By combining quark and antiquark states, the mesons could also be derived.

Quark	I	L_3	S	Q	B	Y=[B + S]
u	$\frac{1}{2}$	$\frac{1}{2}$	0	$\frac{2}{3}$	$\frac{1}{3}$	$\frac{1}{3}$
d	$\frac{1}{2}$	$-\frac{1}{2}$	0	$-\frac{1}{3}$	$\frac{1}{3}$	$\frac{1}{3}$
s	0	0	–1	$-\frac{1}{3}$	$\frac{1}{3}$	$-\frac{2}{3}$

Generation of Baryons from quark combination.

Particles	*Quark*	*S*	*Q*	*B*	*Y*
Protons P	uud	0	1	1	1
Neutron N	udd	0	0	1	1
Lambda Λ°	uds	–1	0	1	0
Sigma Σ^*	uus	–1	1	1	0
Sigma Σ^0	uds	–1	0	1	0
Sigma Σ^-	dds	–1	–1	1	0
$x_i\Xi^\circ$	uss	–2	0	1	–1
$x_i\Xi^-$	dss	–2	–1	1	–1
Ω^-	sss	–3	–1	1	–2
π^+	$u\bar{d}$	0	+1	0	0

The fact that we have three quarks is now a days referred to as the quarks coming, in three flavours. Further refinement says that each quark flavours is met in three colours (red, white and blue) and observed hadrons are all colourless combination of the coloured quark flavours.

Bibliography

Anotola, S.A. : *Advanced Physics,* Longman, London, 1968.

Atherton, J.C. : *Thermal Physics,* Butterworths, London, 1987.

Atiyah, M. : *The Geometry and Physics of Knots,* Cambridge, New York, 1991.

Bemier, G. and Kinet J.M. : *Thermal and Statistical Physics,* C.R.C. Press, Boca Raton, 1981.

Benjamin, W.A. : *Problems in Physics,* Academic Press, New York, 1970.

Bowling, D.J.F. : *Basic Physics,* Chapmann and Hall, London, 1976.

Caro, D. E. and Others : *Modern Physics : An Introduction to Atomic and Nuclear Physics,* Edward Arnold, London, 1985.

Cartwright, H.R. : *Elements of Thermal and Statistical Physics,* Oxford University Press, London, 1966.

Clinton, Case M. : *Physical Principles of Flow in Unsaturated Porous Media,* Oxford, New York, 1994.

Crafts, A.S. : *Phloem Transport in Plants,* W.H. Freeman, San Francisco, 1973.

Douglas, J.S. : *Advanced Guide to Modern Physics,* Pelham Books, London, 1976.

Epstein, E. : *Statistical Physics,* John Wiley and Sons, New York, 1972.

Esau, K. : *Physical Geology*, John Wiley and Sons, New York, 1965.

Evans, L.T. : *Thermal Analysis of Materials*, W.A. Benjamin, California, 1975.

Feynman and Others : *The Feymman Lectures on Physics*, Add. Wesley, London, 1989.

Feynman, R. P. and Others : *Elementary Particles and the Laws of Physics*, Cambridge, New York, 2002.

Ftzler M.E. : *Plant Lectins : Molecular and Biological Aspects*, Academic Press, New York, 1985.

Fujimoto, Minoru : *The Physics of Structural Phase Transitions*, Springer, London, 2000.

Galliard, T. and Mercer, E.I. : *Recent Advances in the Chemistry and Biochemistry of Plant Lipids*, Academic Press, New York, 1975.

Green, J.R. : *Modern Physics*, MacMillan, London, 1911.

Haasen, Peter : *Physical Metallurgy*, Cambridge, New York, 1996.

Halevy, A.H. : *Handbook of Physics*, C.R.C. Press, Boca Raton, 1985.

Hopkins, C.R. : *Structure and Function of Cells*, W.B. Saunders Co., London, 1978.

Hornberger, George M. : *Elements of Physical Hydrology*, Johns Hopkins, New York, 1998.

Jorgensen, T. P. : *The Physics of Golf*, Springer, New York, 1999.

Kirkby, E.A. : *Mathematical Physics*, The University of Leeds, Leeds, 1970.

Kuhn, Karl F. : *Basic Physics*, John Wiley, London, 1996.

Kumar, Arun : *Problems in Physics*, Atlantic Publishers, New Delhi, 2003.

Landshoff, P. and Others : *Essential Quantum Physics*, Cambridge, New York, 2001.

Leo, William R. : *Techniques for Nuclear and Particle Physics Experiments : A How to Approach,* Springer, London, 1994.

Lipson, S. G. and Others : *Optical Physics,* Cambridge, New York, 1995.

Lockett, Keith : *Physics in the Real World,* Cambridge, New York, 1995.

Lui, Lam : *Introduction to Nonlinear Physics,* Springer, New York, 1997.

Main, Lain G. : *Vibrations and Waves in Physics,* Cambridge, New York, 1998.

McKee, H.S. : *University Physics,* Oxford University Press, London, 1962.

Meidner, H. and Sheriff, D.W. : *Water and Plants,* Blackie and Sons, Glasgow, 1976.

Mengel, K. and Kirkby, E.A. : *Thermal and Statistical Physics,* Praeger Publishers, New York, 1974.

Nabapro, F. R. N. and Others : *The Physics of Creep,* T & F, New York, 1995.

Negele, J. W. and Others : *Advances in Nuclear Physics,* Plenum, London, 1994.

Palladin, V.I. : *Physical Metallurgy,* Arihant Publishers, Jaipur, 1988.

Plischke, M. and Others : *Equilibrium Statistical Physics,* World Scientific, New York, 1994.

Rabinowitch, E. : *Thermodynamics and Statistical Physics,* John Wiley, New York, 1969.

Rawn, J.D. : *Encyclopaedia of Modern Physics,* Harper and Row, New York, 1983.

Roberts, J.A. and Tucker, G.A. : *Physics and Chance,* Oxford, New York, 1999.

Rossi, Paolo : *The Birth of Modern Science,* Blackwell, New Jersey, 2001.

Ruhland, W. : *Encyclopaedia of Thermodynamics Physics*, Springer-Verlag, Berlin, 1961.

Sabnis, D.D. and Hart, J.W. : *Encyclopaedia of Stastical Physics*, Springer Verlag, Berlin, 1982.

Sagdeev, R. Z. : *Nonlinear Space Plasma Physics*, AIP, New Jersey, 1993.

Salisbury, F.B. and Ross, C.W. : *Thermal Physics*, Woodworth Publishing Co., California, 1967.

Schein, L. B. : *Electro-Photography and Development Physics*, Springer, London, 1992.

Slatyer, R.O. : *The Physics of Structural Phase Transitions*, Academic Press, New York, 1967.

Srivastava, H.S. : *Elements of Physics*, Rastogi Publications, Meerut, 1990.

Thimann, K.V. : *Physics and the Real World*, CRC Press , Boca Raton, Florida, 1980.

Tolbert, N.E. : *Microbodies—Peroxisomes and Glyoxysomes*, Oxford University Press, London, 1971.

Toraldo, Francia G. : *The Investigation of the Physical World*, Cambridge, New York, 1981.

Verner, J.E. : *Thermal Physics*, Academic Press, New York, 1985.

Vince, Pruce D. : *Elements of Stastical and Thermodynamics Physics*, McGraw Hill, London, 1975.

Walgraif, Daniel : *Introduction to Physics*, McMillan Press, Melbourne, 1969.

Young, H. D. and Others : *University Physics*, Add. Wesley, London, 1996.

Zeevart, J.A. : *Thermal Plasma and New Materials Technology*, McGraw Hill, London, 1976.

Zhukov, L. : *Modern Physics in Twenty first Century, Springer Verlag, Berlin, 1975.*